Special Thank

I am deeply grateful to my parents, whose unwavering support and encouragement, especially during the challenging moments, have shaped me into the person I am today.

They didn't just give me life, they inspired me to live it to the fullest, always urging me to grow, strive, and never settle.

Thank you, Mom and Dad, from the bottom of my heart.

Author: Jonathan Cantin
Copy edited by: Albert Cantin

Mastering CNC and Digital Fabrication
The Ultimate Practical Guide to CNC Routing, Laser, Plasma, Waterjet, 3D Printing, and Welding

ISBN: 1-896369-53-7 // 978-1-896369-53-2

V3

i

TABLE OF CONTENTS

Foreword: Why Making Still Matters   p. xi

Chapter 1: Unlocking Custom Manufacturing   **p. 1**

1.1	What's Custom Manufacturing?	p. 1
1.2	CNC: The Digital Backbone	p. 2
1.3	Brief CNC Evolution	p. 3
1.4	The Custom Manufacturing Workflow	p. 3
1.5	Prototype First, Pretty Later	p. 4
1.6	Common CNC Machines	p. 4
1.7	Specialized CNC Machines in Manufacturing	p. 7
1.8	CNC vs. Traditional: Quick Guide	p. 8
1.9	Design for Reality	p. 9
1.10	UnSuccess Story: Evolution of SBR	p. 10
1.11	Practical Takeaways from the Shop	p. 13
1.12	Chapter 1 Quiz	p. 14
1.13	Answer Key + Explanations	p. 18

Chapter 2: Designing for Real Results   **p. 21**

2.1	What's DFM (Design for Manufacturing)?	p. 21
2.2	Kerf: The Material Loss Line	p. 21
2.3	Nesting: Efficient Material Usage	p. 23
2.4	Planning Fit and Finish: Tolerances Matter	p. 26
2.5	Design with Outcome and Tool in Mind	p. 26
2.6	Working With CNC Machine Constraints	p. 27
2.7	Prepping Files for CNC Success	p. 28
2.8	Success Story: Custom Floor Vents	p. 30
2.9	Practical Takeaways from the Shop	p. 34
2.10	Chapter 2 Quiz	p. 34
2.11	Answer Key + Explanations	p. 38

Chapter 3: Smart Material Selection   **p. 41**

3.1	Why Does Material Selection Matter?	p. 42
3.2	Dealing with Large Suppliers	p. 43
3.3	Material Categories	p. 44

3.4	Wood & Wood Products	p. 45
3.5	Plywood ≠ Plywood	p. 46
3.6	Routers Rule for Thick Plywood Cuts	p. 46
3.7	Plastic Products	p. 47
3.8	Metal Products	p. 48
3.9	Choosing Materials by Application	p. 49
3.10	Success Story: Bone Corian Outdoor Plaques	p. 50
3.11	Practical Takeaways from the Shop	p. 55
3.12	Chapter 3 Quiz	p. 55
3.13	Answer Key + Explanations	p. 59

Chapter 4: CNC Routing That Shapes Ideas — p. 61

4.1	What's a CNC Router?	p. 62
4.2	Why Does CNC Win (Most of the Time)?	p. 62
4.3	Router vs. Spindle	p. 63
4.4	What Can CNC Routing Do for YOU?	p. 63
4.5	Tooling 101: Bits & End Mills	p. 64
4.6	Feed Rates & Speeds: Simplified	p. 66
4.7	Chip Load ≠ Fancy Nacho	p. 66
4.8	Workholding: Keeping Material in Place	p. 67
4.9	Common Problems & Fixes	p. 68
4.10	Material Considerations for Routing	p. 68
4.11	Finishing Impacts Strategy	p. 69
4.12	Toolpath Strategy Basics	p. 69
4.13	Tool Life Factors	p. 70
4.14	Quality Tools Are Cheaper Than Cheap Ones	p. 70
4.15	The High Cost of Dull Tools	p. 71
4.16	Success Story: Oversized Oak Door Engraving	p. 72
4.17	Practical Takeaways from the Shop	p. 76
4.18	Chapter 4 Quiz	p. 77
4.19	Answer Key + Explanations	p. 81

Chapter 5: CNC 3D Printing Builds Dreams — p. 83

5.1	What's 3D Printing, Really?	p. 84
5.2	Choosing the Right 3D Printing Tech	p. 85
5.3	Common 3D Printing Materials	p. 85
5.4	• FDM Material Options	p. 86

5.5	• Resin Material Options	p. 87
5.6	• Metal Material Options	p. 88
5.7	3D Printer Troubleshooting	p. 89
5.8	Bridging the Digital and Real Worlds	p. 90
5.9	Strengths and Weaknesses of 3D Printing	p. 91
5.10	When to 3D Print vs. When to CNC	p. 93
5.11	Design Tips for 3D Printing	p. 95
5.12	Real-World Applications	p. 96
5.13	3D Printing: Slow Bakes and Sudden Breaks	p. 97
5.14	Future Trends in 3D Printing	p. 97
5.15	Success Story: Multi-CNC Corporate Awards	p. 98
5.16	Practical Takeaways from the Shop	p. 104
5.17	Chapter 5 Quiz	p. 105
5.18	Answer Key + Explanations	p. 109

Chapter 6: CNC Laser Precision and Possibility **p. 111**

6.1	What's a CNC Laser?	p. 112
6.2	CO_2 vs. Fiber Laser, Power and Precision	p. 112
6.3	Gantry vs. Galvo, Speed and Scale	p. 113
6.4	Brief Laser Evolution	p. 114
6.5	Laser Applications Breakdown	p. 114
6.6	Air Assist & Exhaust Systems	p. 115
6.7	Material Compatibility Guide	p. 116
6.8	Laser Parameters: Speed, Power, Frequency	p. 116
6.9	Low-Wattage vs. High-Power Fiber Lasers	p. 117
6.10	Focus & Lensing	p. 117
6.11	Support Your Manufacturer	p. 118
6.12	Common Laser Issues & Fixes	p. 119
6.13	Design Prep Checklist	p. 119
6.14	Summary of Laser Types	p. 119
6.15	Success Story: The GitHub Arctic Code Vault	p. 120
6.16	Practical Takeaways from the Shop	p. 123
6.17	Chapter 6 Quiz	p. 124
6.18	Answer Key + Explanations	p. 125

Chapter 7: CNC Plasma Ionized Firepower **p. 129**

7.1	What's Plasma Cutting?	p. 129

7.2	Plasma Cutting: Manual vs. CNC vs High-Def	p. 130
7.3	When Steel Gets Tough, Plasma Gets Going	p. 131
7.4	Plasma Cutting Applications	p. 132
7.5	Material Compatibility Guide	p. 133
7.6	Key Plasma Cutting Parameters	p. 133
7.7	Manual vs. CNC vs. Oxy-Acetylene Cutting	p. 135
7.8	Success Story: Cottage Sign Fabrication	p. 136
7.9	Practical Takeaways from the Shop	p. 139
7.10	Chapter 7 Quiz	p. 139
7.11	Answer Key + Explanations	p. 143

Chapter 8: Welding for Fabrication Success | **p. 145**

8.1	What's Welding?	p. 146
8.2	Welding is Everywhere	p. 146
8.3	Core Welding Processes	p. 147
8.4	MCAW vs GMAW	p. 147
8.5	Welding Without Gas Tanks	p. 149
8.6	Brief Evolution of Welding	p. 150
8.7	Shielded Metal Arc Welding (SMAW)	p. 150
8.8	Commonly Used SMAW Rods	p. 151

Want perfect square welds? Use corner clamps.

8.9   Understanding SMAW Rod Classifications   p. 151
8.10   Welding Tips & Tricks (FCAW-S + SMAW)   p. 152
8.11   Pipe vs. Tube   p. 153
8.12   Tube & Pipe Welding Cost Saver   p. 154
8.13   Cold Cutting   p. 154
8.14   Welding Symbols   p. 155
8.15   Success Story: Custom Fire Pit   p. 156
8.16   Practical Takeaways from the Shop   p. 158
8.17   Chapter 8 Quiz   p. 159
8.18   Answer Key + Explanations   p. 163

Chapter 9: CNC Nesting Optimization Tactics   **p. 164**

9.1   What's Nesting?   p. 166
9.2   CNC Nesting Applies To All CNC Systems   p. 166
9.3   Why Do Most Shops Go Hybrid?   p. 167
9.4   Key Nesting Considerations   p. 167
9.5   Nesting in Workflow   p. 168
9.6   Nesting = More Margin   p. 168
9.7   Nesting Across Materials   p. 169

Think you have enough clamps? You don't. Buy three more.

9.8	Success Story: Plywood Big Wheels	p. 170
9.9	Practical Takeaways from the Shop	p. 173
9.10	Chapter 9 Quiz	p. 174
9.11	Answer Key + Explanations	p. 177

Chapter 10: Mastering the CNC Superheroes — **p. 179**

10.1	The CNC Avengers: Choosing the Right Hero	p. 180
10.2	The CNC Avengers: Ultimate Synergies	p. 182
10.3	The CNC Avengers: Expanding the Team	p. 183
10.4	Success Story: Christmas Ornaments	p. 184
10.5	Practical Takeaways from the Shop	p. 187
10.6	Chapter 10 Quiz	p. 188
10.7	Answer Key + Explanations	p. 193

Chapter 11: CNC Robotics and Automation — **p. 195**

11.1	Robotics in CNC: Beyond Welding	p. 196
11.2	What's a Robotic Arm?	p. 197
11.3	When Does Automation Make Sense?	p. 197
11.4	My Shop's CNC Router Evolution	p. 198
11.5	Lights-Out Manufacturing	p. 199
11.6	Comparing Old vs. New Automation Models	p. 200
11.7	Modern Tools for Flexible Automation	p. 200
11.8	Emerging Technologies in CNC	p. 201
11.9	Practical Takeaways from the Shop	p. 202
11.10	Chapter 11 Quiz	p. 202
11.11	Answer Key + Explanations	p. 207

Chapter 12: From Prototyping to Production — **p. 209**

12.1	What Makes a Prototype?	p. 209
12.2	The Prototyping Process	p. 210
12.3	Lessons from Scrap Heap University	p. 212
12.4	Prototype to Production Workflow	p. 215
12.5	Nested for Speed	p. 216
12.6	Scaling Work	p. 216
12.7	UnSuccess Story: The Drais BMT Journey	p. 217
12.8	Practical Takeaways from the Shop	p. 221

12.9   Chapter 12 Quiz   p. 221
12.10 Answer Key + Explanations   p. 225

Chapter 13: Finishing Techniques That Matter   p. 227

13.1   Surface Finishing   p. 228
13.2   Marking: Identity, Instruction, and Branding   p. 228
13.3   Assembly Considerations   p. 230
13.4   Don't Skip Finishing in Your Quote   p. 231
13.5   Cleaning and Quality Control   p. 232
13.6   Cleaning Tools to Keep on Hand   p. 234
13.7   Marking vs Engraving   p. 234
13.8   Success Story: Custom Plywood Boxes   p. 235
13.9   Practical Takeaways from the Shop   p. 238
13.10 Chapter 13 Quiz   p. 239
13.11 Answer Key + Explanations   p. 243

Chapter 14: Packaging and Delivery Strategy   p. 245

14.1   Packaging   p. 246
14.2   Clear Documentation: Labels and Instructions p. 246
14.3   Delivery Methods: Speed and Costs   p. 247
14.4   Packaging for International Shipping   p. 247
14.5   Environmental Considerations in Packaging   p. 249
14.6   Client Handoff: More Than a Transaction   p. 250
14.7   Packing as Part of Your Brand   p. 250
14.8   Success Story: Custom Foam Inserts   p. 251
14.9   Practical Takeaways from the Shop   p. 254
14.10 Chapter 14 Quiz   p. 254
14.11 Answer Key + Explanations   p. 259

Chapter 15: Future-Proofing CNC Workflow   p. 261

15.1   The Rise of Cobots & Automation   p. 262
15.2   AI in Custom Manufacturing   p. 263
15.3   Next-Gen Tech: What's on the Horizon?   p. 263
15.4   Stay Curious: Experiment to Stay Ahead   p. 266
15.5   Sustainability & Circular Design   p. 267
15.6   Why Does Sustainability = Smart Business?   p. 267
15.7   Off-Planet Manufacturing   p. 268

15.8   Practical Takeaways from the Shop   p. 269
15.9   Chapter 15 Quiz   p. 269
15.10 Answer Key + Explanations   p. 274

Final Words: The Build Never Really Ends   p. 276

Q&A: Your CNC Questions Answered   p. 278

Because One Book Can't Cover Everything   p. 295

Appendices: Deeper Dives and Handy Extras

Appendix A: Imperial–Metric Conversion Sheet   p. 296
Appendix B: Metal Gauge vs. Thickness   p. 297
Appendix C: Material Selection Guide   p. 298
Appendix D: Chip Load Explained   p. 302
Appendix E: Adhesives   p. 304
Appendix F: It Looked Great in the Slicer...   p. 306
Appendix G: Historical Concise Dives   p. 316
    • Origins of CNC   p. 316
    • Welding Through the Ages   p. 317
    • Lasers   p. 318
Appendix H: Glossary of Terms   p. 320

Other Books by Jonathan Cantin   p. 350

CNC Waterjets: when stainless steel says "nice try" to hand tools.

AI-powered robots running CNCs, so, where do humans fit in?

Foreword
Why Making Still Matters
Hands-On Skills in a Digital-First World

It's been a decade since I last wrote a book, and both my business and the custom manufacturing world have evolved drastically. What started in a two-car garage is now a full-blown custom fabrication shop, shaped by everything from global chaos to parenthood.

CNCROi.com quickly pivoted to B2B clients after realizing personalized keychains wouldn't cover the bills. I stopped selling digital files on CNCKing.com but kept the site as a tribute to where it all began. It was fun while it lasted.

Since then, I've gone from a ShopBot Desktop to running a Thermwood M42-55DT router, a Speedy 400 flexx laser, a 4×8 ft CNC plasma cutter, and even went to welding school to cross it off my bucket list. Somewhere in there, I also became a dad, and everything changed.

This book isn't an encyclopedia, it's a snapshot of the lessons, skills, and chaos behind running a custom shop, one of the toughest businesses to get into, let alone stay in. It shares just a slice of what it takes to make it all work efficiently and profitably.

The CNC world moves fast, and this is where I'm personally at in early 2025. I wrote this for my son, Simon, not to show him how to push a button, but how to build the machine most people never realize is running the show.

It also came from surviving two years in Niagara College's full-time Welding Technician program... at 48 years old. The textbooks were bone-dry, with barely a nod to CNC or its role in modern manufacturing, even in areas where welding plays a part.

Fortunately, the instructors were seasoned pros who brought real-world energy to the classroom. I learned a lot, but the CNC gap was glaring, and I knew it was mine to fill, given my decades of experience. ***Hence, this book!***

Jonathan Cantin, The CNC King
(and founder of CNCROi.com, where these lessons are forged daily)

Blame the club, it isn't the CNC laser-engraved divot.

The custom gasket, unsung hero of non-explosions.

Chapter 1
Unlocking Custom Manufacturing
Turn Ideas Into Tangible, Real-World Creations

Every project is a fresh challenge, whether you're making a sign that stops traffic, a prototype that actually works, or a panel that's more sculpture than structure. Each one pushes you to unlock custom manufacturing in a new way, turning ideas into tangible, real-world creations, sometimes messy, often surprising, but always unique.

You're juggling design constraints, tool limits, and material quirks, often all at once. It's less about pressing buttons and more about thinking critically, and adapting quickly toward smart, efficient solutions that solve real problems.

This chapter explores the hands-on reality of fabrication, when to let machines take over, and when a manual tool (and a bit of stubbornness) wins the day.

Choosing the right direction is a winding process guided by passion and just a little luck.

💬 *It's not about mass production, it's value.*

1.1 What's Custom Manufacturing?

At its core, custom manufacturing is the art of building something that doesn't already exist yet, at least not in that specific size, material, shape, or style.

It's about translating ideas into physical objects that meet specific needs, whether for a client, a company, or your own wild project that came to you in a vision.

As we go from pixels to dust, we'll see how CNC tech complements traditional craftsmanship, and how choosing the right mix of methods can mean the difference between a frustrating job and a fantastic result.

👎 *Skipping the Why*

CNCROi.com started in my parents' garage in 2014.

1.2 CNC: The Digital Backbone

CNC stands for Computer Numerical Control. It's what tells machines exactly where and how to move.

CNC goes far beyond routers, it drives lasers, waterjets, plasma tables, mills, and more. Even your car is technically a CNC machine these days.

It's precision at scale, even if that scale means producing just one or a handful of units. Repeatability and accuracy are built into the process.

But CNC isn't always the answer.

While these machines bring unparalleled precision and control, traditional tools like welders, saws, and drills still shine in certain scenarios, especially when agility, instinct, or on-the-fly adjustments are needed.

✐ Design for the process, not just the looks.

📖 1.3 Brief CNC Evolution (see "Origins of CNC: From Punch Cards to AI" in Appendix)

Automation began in the 1940s with punch cards and servos. The 1960s brought G-code and CNC precision. In the 1980s, microchips and CAD/CAM boosted adoption. By the 2000s, desktop CNC empowered makers. Today, AI, cloud, and cobots drive smart manufacturing.

1.4 The Custom Manufacturing Workflow

The custom manufacturing workflow is like creating a fine meal, except you're using machines instead of a stove, and your ingredients are metal, wood, or plastic. Sometimes with a dash of unknown.

It's not about rigid formulas, but about understanding the balance of process, timing, and material, like a seasoned chef who relies on tools, instincts, experience, and the subtle feel of each ingredient in the moment.

My first shop after graduating from my parents' garage in 2017.

1. Consultation & Design Brief: First, clarify the end goal, what's the purpose, environment, longevity, and aesthetics? Ask the questions the client didn't even know they should be asking. "Do you want this to survive an earthquake or just look pretty on Instagram?"

2. Design (CAD): Turn those ideas into a digital reality, think of it as sketching, but with more buttons, fewer coffee stains, and a lot more CTRL+Z. Don't get lost in the pretty pictures. First, nail down the budget and timeline; otherwise, you'll be designing a mansion on a cardboard-box budget.

3. Material Selection: No material does it all. Choose based on strength, cost, and compatibility. Don't pick the cool, shiny material just because it looks good... unless your project is an online filter come to life. Then go wild.

4. Machine Selection: Match the design to the right tool, or mix of tools. Remember, every machine has its quirks. Sometimes the best option isn't CNC at all, maybe it's a welder or an old-school table saw that's been hanging around since your grandpa's day. You know, the one with character that doubles as a safety hazard.

5. Cut It Out: Time to make the magic happen (and possibly create some chaos in the process). Run the job with precision, efficiency, and safety. And don't forget to share those cool process photos with the client, nothing says "we're doing this right" like an action shot of a laser beam slicing through steel.

6. Ship It Out: Pack it up and ship it out. Whether it's one part or a thousand, make sure it's perfect before it leaves your hands. Nothing's worse than a last-minute "Oh, we forgot to add that one crucial detail" after the delivery truck pulls away.

And that, folks, is how you go from a sketch to a shiny, functional product that'll make your client's dreams come true, unless, of course, they forgot to provide a proper design brief. Then all bets are off.

⬇ 1.5 Prototype First, Pretty Later

In custom manufacturing, version one (or 1 000) doesn't need to be perfect, it just needs to work. Too many people chase polish before proof. If it can't be sketched on a napkin, it's probably overdesigned.

Ask yourself: Is form or function more important? Can I simplify to save time? Am I designing for the shop or just the screen? Solve first, refine later.

1.6 Common CNC Machines in Manufacturing

This table highlights the CNC machines I use frequently at *CNCROi.com*. Knowing their strengths and limitations is an important step toward picking the best one for a project.

Machine	Best For	Strength	Limitations
CNC 3D Printers	Prototyping and producing complex geometries in plastics and metals	Design freedom and low waste	Print failures, material limitations, and slower production speed
CNC Lasers (CO$_2$ & Fiber)	Cutting, engraving, and etching wood, acrylic, thin metals	Detail and speed	Flammable materials and heat discoloration
CNC Plasma Cutters	Cutting conductive metals like steel and aluminum	Speed and power, especially for thick sheets	Wide kerf and dross (molten metal spatter)
CNC Routers	Wood, plastic, soft metals	Versatility with 3D shaping	Tool chatter, especially in hard materials.

Machine	Best For	Strength	Limitations
CNC Waterjets	Virtually anything that won't absorb water, including stone and composites	Cold cutting = no warping from heat-affected zones (HAZs)	Abrasive cleanup and cost per cut can be high.

⚠️ *The welder isn't CNC, but it's the crucial sidekick, turning cut metal into solid assemblies and even supporting non-metal builds.*

In some setups, welders pair with CNC cobots, collaborative robots, to speed up repetitive welding while maintaining consistency.

Welding is like metal glue, only far stronger and hotter, fusing pieces at a molecular level to create lasting strength.

First welding table, March 18, 2018. No experience, all optimism.

1.7 Specialized CNC Machines in Manufacturing

These machines aren't in my current setup, but trust me, you're always one job away from needing one of these in your shop. I've pivoted more times than I'd like to admit, usually right after saying, "I'll never *need* that."

Machine	Summary Use Case	Why Not Covered in Depth
CNC EDM	Precision for hardened metals via electrical discharge	Highly specialized with almost no kerf, like a precision surgeon with a laser instead of a scalpel.
CNC Lathe	Round or cylindrical parts	Less relevant for flat sheet work, but opens up a whole new world, like finding a hidden level in a video game.
CNC Mill	Multi-axis metal shaping	Covered only tangentially, since it's essentially a CNC router for metal, like a CNC router's cooler, tougher metal chewing sibling.
CNC Punch Press	Stamping holes or shapes in sheet metal	Great for high-volume, repetitive operations, but less versatile for complex shapes.
CNC Stamping	High-speed mass production of metal parts	Primarily used for high-volume, identical parts, ideal for large runs, but not for the custom, one-off jobs I focus on.
CNC Bending Machine	Bending metal sheets or tubes to precise angles	Extensively used in custom manufacturing, and I'm eyeing them as my metal division grows.

1.8 CNC vs. Traditional: Quick Guide

Sometimes it's not about what you can do, but what you should do. Knowing which side of this chart your project leans on helps you stay efficient and avoid overengineering.

CNCs aren't always the best tool for the job!

Factor	CNC Fabrication	Traditional Fabrication (Manual)
Precision & Repeatability	High precision, ideal for complex, detailed designs	Varies by skill and consistency
Speed (Production)	Faster for complex or repeated parts	Quick for one-offs and fixes, slower for complexity
Skill Requirement	Requires CNC programming and setup knowledge	Relies on experience and craftsmanship
Cost (Short Runs)	High setup, low per-part cost	Low setup, high per-part labor
Flexibility	Limited by machine design and programming	High, adjustments can be made on the fly
Ideal Use Case	Mass production, precise or complex parts	Quick fixes, prototyping, personal touch
Setup Time	Longer (software + fixturing)	Shorter, ready to go after you plug it in

CAD's real power is simulating, not guessing.

1.9 Design for Reality

CAD files are like that overachieving student who aced the test but skipped the homework, great on screen, but clueless on the shop floor.

Machines, on the other hand, deal with the real-world mess of clamps, kerf, and heat distortion all show up like uninvited guests. CNCs don't know what's around them, it's up to the operator to turn clean virtual designs into clean cuts in the real world to the spec required.

That flawless CAD model? It might collide with clamps, ignore kerf, or warp under heat, because the machine blindly follows lines of code, not reality.

Design with your machines and their quirks in mind, no one likes surprises when it's time to cut, weld, or carve.

Software is catching up. Vectric, for example, now lets you set 'no-go zones' to prevent your machine from crashing into clamps on its way back to 0/0/0 (the XYZ origin of the CNC machine), a small fix that can prevent a big mess.

▢ 1.10 UnSuccess Story
Evolution of SandboxRanch.com

SandboxRanch.com was originally built around in-person engagement at craft shows but had to quickly adapt when the COVID-19 pandemic shut those events down.

As customers shifted online, the company pivoted to survive, a clear example of how flexibility is key in changing market conditions.

All set up for an arena craft show.

Process Breakdown

1. Initial Business Model and Growth

SandboxRanch.com thrived by attending artisan markets and craft shows, selling directly to customers. These face-to-face interactions fostered brand loyalty and created consistent revenue streams. In contrast, *CNCROi.com* operates on a B2B model with minimal in-person interaction.

2. Pandemic Disruption

With in-person events canceled, the core of the business collapsed. Attempts to transition to online sales revealed new challenges: maintenance costs for e-commerce platforms and the ongoing investment required for digital marketing.

Meanwhile, fewer customers were searching for non-essential goods online, shrinking a once-eager customer base. It was like being stuck between two shrinking markets without a way out.

3. Challenges and Solutions

Rising Overhead Costs:
Transitioning to an online-only business model brought a wave of unexpected expenses, website maintenance, digital advertising, platform fees, and more complex logistics. Unlike a craft booth and cashbox, e-commerce demanded constant investment, just to stay visible, let alone competitive.

Shrinking Market Access:
Without the foot traffic and impulse buying of live events, SandboxRanch.com faced a steep drop in casual sales. To adapt, the brand explored new product lines that served different niches and tested various online marketplaces to expand reach. It wasn't just about moving online, it was about reinventing how and where customers discovered the products.

Reassessing the Model:
Rather than continue chasing one-off, small-ticket sales, the business pivoted toward creating bundles and customizable kits, encouraging higher order values. This approach reduced the marketing effort per transaction, and helped balance out fulfillment costs.

4. Rebuilding and Resilience

The path forward wasn't clear-cut. Progress didn't follow a straight line, it zigzagged through trial, error, and relentless refinement. While online-only sales never fully replicated the in-person show success, they unlocked a new kind of opportunity: collaboration.

CNCROi.com began helping customers who wanted to enhance their own product lines with customized, made-to-order components.

That pivot inspired others to reimagine what their businesses could be when supported by tailored manufacturing services. Just because you have a booming business one year, doesn't mean it will be the same in the next. Fail quick and often, don't forget the lessons.

Custom banks ended-up costing more than what goes in them.

Things I Had to Learn the Hard Way

1. Adaptability is Key

When unexpected disruptions hit, like a global pandemic, the ability to change course quickly becomes more important than sticking rigidly to the original plan. Flexibility keeps momentum going when certainty disappears.

2. Diversify Sales Channels

Relying solely on one revenue stream (like craft shows) creates vulnerability. Multiple sales avenues, online stores, custom orders, B2B services, add flexibility, reach, and long-term security.

3. Stay Alert to Change

Markets evolve. Keeping an eye on trends, customer behavior, and emerging tech allows businesses to act early and stay ahead rather than scramble to catch up when it's already too late.

Another One Off the Table

This case study is a testament to entrepreneurial adaptability.

What started as a local side hustle grew into a thriving brand, until the pandemic brought craft shows, and momentum, to a halt. With its main sales channel gone, the business had to rethink everything.

Instead of giving up, it pivoted. New product lines, digital platforms, and collaborations replaced foot traffic.

The transition wasn't easy, but it revealed something deeper: success isn't about avoiding disruption, it's about evolving with it, without losing the passion that started it all. It isn't fun to be forced to pivot into something else.

SandboxRanch.com used challenge as a catalyst for reinvention.

🎖 1.11 Practical Takeaways from the Shop

Don't dive into CNC headfirst: Start by defining the actual problem, machines aren't mind readers, they're obedient pixel chompers. Overdesign in CAD, and your CNC will faithfully follow, right into a mess.

Choosing the wrong machine for your material: It's like putting snow tires on a racecar, technically possible, but wildly unhelpful. Know your materials. Know your tools. And don't assume "faster" means "better."

Treat scrap like a rehearsal dinner: Low pressure, low cost, and the perfect time to screw up before the big show. Test, tweak, and only then, commit to that pristine slab of walnut or sheet of brushed aluminum.

Don't let your CNC get all the glory: Sometimes the hero is a humble clamp, or your own two hands. CNC is a partner, not a replacement, think Batman and Robin, not Batman versus the Joker.

❌ 1.12 Chapter 1 Quiz

Time to see if you've been paying attention or just nodding off!

1. What's the primary focus of custom manufacturing?
    A) Mass production at scale
    B) High-speed automation for repeat parts
    C) Reducing the cost of consumer goods
    D) Tailored, low-volume products for specific needs

2. What does CNC stand for?
    A) Controlled Numerical Coding
    B) Computer Numerical Control
    C) Calibrated Neutral Cutting
    D) Computerized Notch Calibration

3. Which of the following is NOT a CNC-powered tool?
    A) CNC waterjet
    B) CNC plasma cutter
    C) Manual Drill Press
    D) CNC laser cutter

4. What's a common rookie mistake in manufacturing?
    A) Jumping to the machine too early
    B) Overuse of manual tools
    C) Skipping CAD design
    D) Choosing expensive materials

5. Why is "designing for the process" important?
    A) It speeds up CAD rendering.
    B) It ensures aesthetic appeal.
    C) It reduces the number of revisions.
    D) It aligns the design with machine limitations.

6. Which CNC tool avoids a heat-affected zone (HAZ)?
    A) CNC plasma cutter
    B) CNC laser
    C) CNC waterjet
    D) CNC router

7. Which factor best favors manual tools over CNCs?
    A) High part repeatability
    B) One-off prototyping with quick changes
    C) Complex 3D geometries
    D) Engraving on flat wood panels

8. In a CNC-driven shop, what role does the welder play?
    A) Joins metal parts made by CNC machines.
    B) Serves as a CNC-integrated marking tool.
    C) Adds decorative finishes.
    D) Performs outdated fabrication methods.

9. What follows material selection in manufacturing?
    A) Shipping
    B) Machine selection
    C) CAD design
    D) Consultation

10. Why might a project not use CNC tools?
    A) CNC machines lack accuracy.
    B) Manual tools are always faster.
    C) The job demands on-the-fly adaptability.
    D) CNC machines cannot cut metal.

11. What material can't a low-powered CNC laser cut well?
    A) Acrylic
    B) Paper
    C) Birch plywood
    D) Thick steel

12. Why does kerf matter in CNC cutting?
    A) It determines the machine's speed.
    B) It affects the tool's lifespan.
    C) It prevents tool overheating.
    D) It affects how precisely parts are cut.

13. Why did SandboxRanch.com pivot?
    A) A factory fire
    B) Increased taxes
    C) The COVID-19 pandemic
    D) Supply chain automation

14. What's a key lesson from SandboxRanch.com's journey?
    A) Adaptability is key during market shifts.
    B) Stick to the original business plan.
    C) Avoid e-commerce.
    D) Focus only on local markets.

15. What is the "Cut It Out" workflow stage for?
    A) Client delivery
    B) CAD modeling
    C) Material machining or processing
    D) Tool maintenance

16. How does CNC outperform manual production?
    A) Spontaneous creativity
    B) Faster, more precise production of complex parts
    C) Lower upfront setup
    D) Less training required

17. What's a disadvantage of CNC plasma cutting?
    A) Slow speed
    B) Requires water for cooling
    C) Wide kerf and messy molten spatter
    D) Not suitable for metal

18. What does "Custom = Creative Constraint" mean?
    A) Always make it look creative.
    B) Design freely without limitations.
    C) Use available materials for faster prototyping.
    D) Work with, not against, machine constraints.

19. What's true about CAD designs?
    A) They always translate perfectly to production.
    B) They often ignore real-world machine issues.
    C) They eliminate the need for prototyping.
    D) They're best created by AI.

20. Why include both CNC and manual tools in this book?
    A) Both have strengths in different situations.
    B) To confuse beginners less.
    C) To sell more equipment.
    D) Manual tools are cheaper.

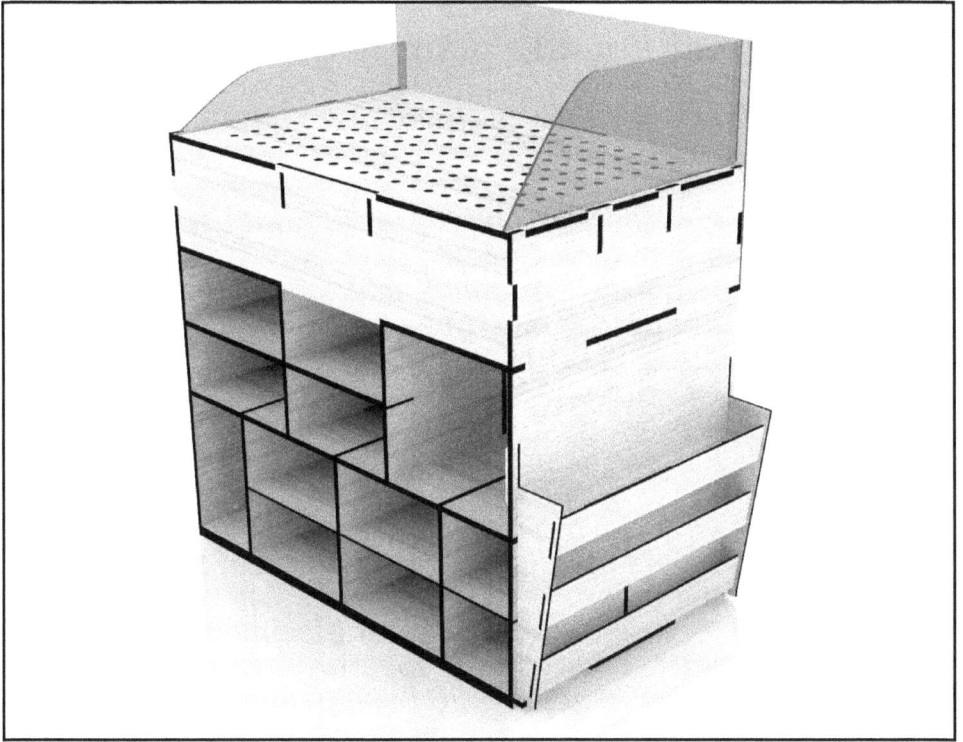

Most projects begin life as a 3D model.

My first CNC project, made on a trusty ShopBot Desktop.

✖ 1.13 Answer Key + Explanations

Here's the answer key, let's see if you were on the money... or way off the mark!

1D – Custom manufacturing focuses on unique, often low-volume, tailored parts.
2B – CNC stands for Computer Numerical Control, the system that guides machine operations.
3C – A handheld drill is not CNC-controlled. The others are all CNC-driven tools.
4A – Beginners often rush into machining without fully understanding the problem or end goal.
5D – Designing for the process ensures that the job is feasible with the tools and materials available.
6C – CNC waterjets use cold cutting, which avoids heat-related distortion.
7B – Manual tools excel in quick, flexible one-offs where agility matters more than repeatability.
8A – Welding complements CNC by allowing assembly and joining beyond cutting or engraving.
9B – After selecting a material, the next step is choosing the right machine to process it.
10C – Manual tools are better for projects needing flexibility and real-time decision-making.
11D – Thick steel is typically too dense and reflective for standard CNC laser cutters.
12D – Kerf is the width of material removed during cutting and affects part sizing.
13C – The pandemic shut down in-person craft shows, disrupting the original business model.
14A – The key takeaway is adaptability, being able to pivot during unexpected changes.
15C – The 'Cut It Out' stage is where actual fabrication, cutting, engraving, etc., occurs.
16B – CNC machines excel at speed and consistency, especially with intricate or repeated parts.
17C – Plasma cutting can be messy, with a wide kerf and molten metal (dross) buildup.
18D – The phrase encourages designing within real-world constraints of tools and materials.
19B – CAD models often ignore shop-floor issues like clamping, kerf, and distortion.
20A – Each method, manual and CNC, has its own advantages, depending on the job.

Lasers: Because sawdust is overrated.

Once a tree. Now a statement.

Rooted in style, branching into greatness. Custom Oak badges.

Custom badges: For when 'Hey you' just isn't good enough.

Chapter 2
Designing for Real Results
Make Every Cut Count from the Start

Designing for real results means more than making something look good, it's about making every cut count. Buildable design is its own skill, and where many good ideas go sideways.

This chapter covers how to design around tool limits, material behavior, and assembly, so your projects not only look right, they work. Your CAD file might say "perfect," but the CNC router might answer with smoke and a scorched edge if you aren't paying attention.

💬 *Designing it is fun. Making it? That's where you find out if your vision was a masterpiece or a disaster.*

2.1 What's DFM (Design for Manufacturing)?

Design for Manufacturing (DFM) is the practice of creating products that are optimized for the actual fabrication process. It's where creativity meets constraint.

You're not just imagining something cool, you're ensuring it can survive cutting, engraving, bending, bonding, and assembly to make it real.

👎 *Designing for the End Result, Forgetting the Process*

2.2 Kerf: The Material Loss Line

Kerf refers to the material removed by the cutting tool, whether it's a laser beam, router bit, or another CNC tool.

The size of the kerf directly impacts how parts fit together:

- **Too tight:** Parts may not fit.
- **Too loose:** Parts may rattle or have excess play.

CNC Laser
- **Kerf Width:** Approximately 0.1–0.3 mm (≈ 1/256 to 1/96")
- **Edge Characteristics:** Clean, sharp edges with minimal heat-affected zone
- **Cut Profile:** Straight, precise cuts with negligible taper

CNC Plasma
- **Kerf Width:** Approximately 1.0–3.8 mm (≈ 1/25–3/20")
- **Edge Characteristics:** Wider cuts with a noticeable taper due to the arc's shape
- **Cut Profile:** Edges may appear slightly beveled, especially on thicker materials

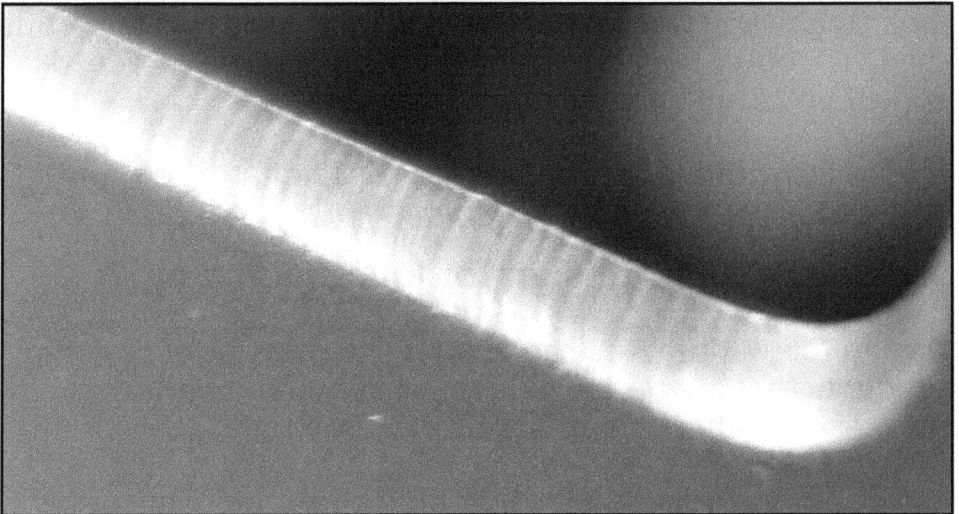

CNC Plasma Cut 12 ga hot-rolled mild steel: Notice the taper.

CNC Router
- **Kerf Width:** Depends on bit size; typically 3+ mm (1/8"+)
- **Edge Characteristics:** Straight cuts with potential tool marks; kerf equals the cutting bit's diameter
- **Cut Profile:** Consistent width throughout, determined by the tool's diameter

CNC Waterjet
- **Kerf Width:** Approximately 0.2–1.0 mm (≈ 1/128–1/32")
- **Edge Characteristics:** Smooth, precise cuts without heat distortion
- **Cut Profile:** Precise, low-taper cuts, ideal for fine details and delicate materials

2.3 Nesting: Efficient Material Usage

Good Nesting
In a good nesting example, parts are placed efficiently, minimizing gaps and maximizing material use.

- Parts are tightly packed with minimal gaps.
- Maximizes material with shared cuts.
- Uses an efficient cutting sequence to reduce setup time.

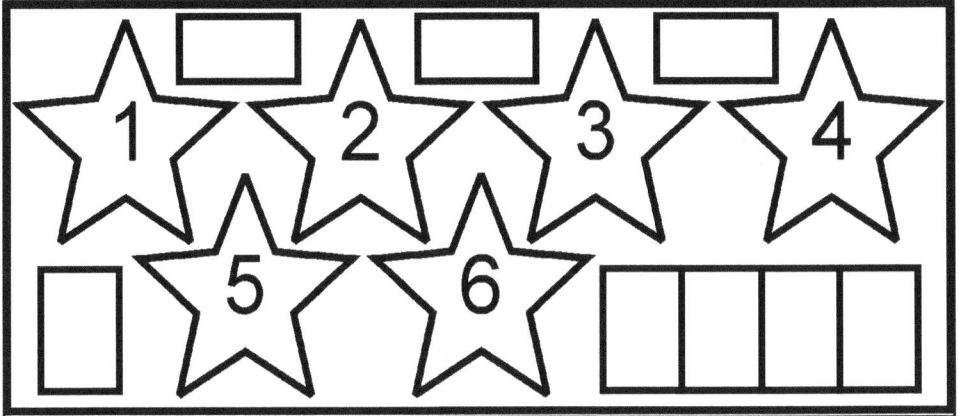

Bad Nesting
In a bad nesting example, parts are poorly arranged, leaving large gaps of unused material that could have been better utilized.

- Large unused gaps between parts.
- Wastes material that could be used for other parts.
- Inefficient cutting sequence that requires extra moves for tool changes.

Thick metal giving you attitude? Waterjet it into submission!

Good nesting maximizes material usage and minimizes waste, making production faster, cleaner, and more cost-effective, giving you higher profit margins!

Proper nesting is crucial in both small- and large-scale manufacturing, ensuring the best use of resources, including time and money.

Tolerance Fit

Always specify tolerances in your CAD files, especially for parts that interact. "Just right" in CAD might end up "just stuck" in reality. Getting the right fit is critical.

That little ± sign buried in customer drawings? Turns out it's not optional, and probably tighter than necessary.

Higher tolerances mean bigger headaches, and bigger bills.

🖉 *Test tolerances on scrap before the full run.*

Toolpaths and Offsets: Impacts on Accuracy

In CNC design, understanding toolpaths and offset paths is key to accurate cuts and proper part fit.

A toolpath is the exact route the cutting tool follows, straight, curved, or any shape matching the part's geometry.

Basic linear toolpath
Start --> |------------------------------------| End

An offset shifts the toolpath to account for tool diameter or kerf, ensuring precision and proper fit. This is especially critical in CNC routing and plasma cutting, where tool size heavily impacts cut accuracy, even a small miscalculation can lead to poor fits or wasted material.

Toolpaths don't exist in isolation, they work in tandem with kerf and offsets to shape every result, from the tightest notch to the cleanest edge.

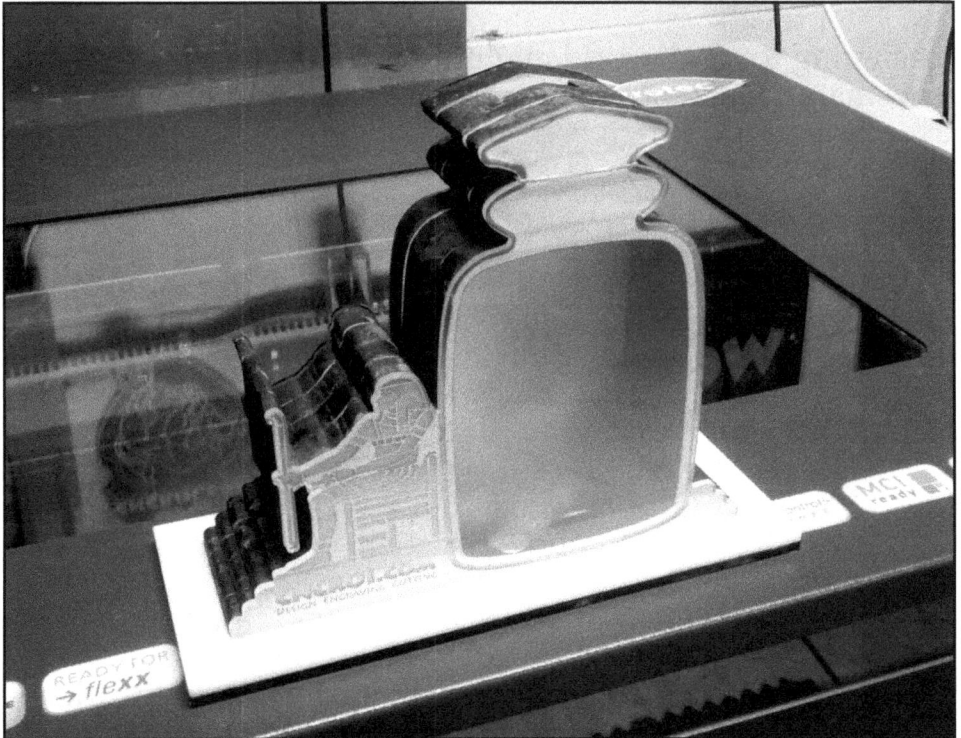

Master toolpathing, bend reality, not your endmill!

⏬ 2.4 Planning Fit and Finish: Tolerances Matter

Let's say you're laser cutting a tab-and-slot box out of 6 mm (1/4") plywood. If you design the slots to be exactly 6 mm wide and your kerf is 0.2 mm (1/128"), guess what? Your tabs won't fit, they'll hit resistance and stop short, no matter how much you push.

They'll be either too tight or too loose, depending on calibration. It's always cheaper to adjust mistakes on scrap than on your final project.

2.5 Design with Outcome and Tool in Mind

In custom manufacturing, imagination isn't the bottleneck, reality is. Unless you have some superhuman insight into the world of physics, you'll always be constrained by them one way or another.

A beautiful design on your screen means nothing if it can't be fabricated cleanly, assembled easily, or repeated reliably.

Designing for real-world manufacturing means anticipating how materials behave, how machines cut, and how humans assemble them, which is the hardest guess to overcome.

Hard Truth

Skip this mindset, and your 'perfect' design might return scorched, skewed, or stuck together like IKEA regret, all sharp edges and zero alignment.

Think like a fabricator:
- How is this cut?
- What's holding this together?
- Will this work in plywood, steel, or acrylic, or just in your head?
- Can someone build this without guessing?

Design is not just what it looks like, it's how well it works when gravity, kerf, and clamps get involved, whether you're on your first prototype or your fiftieth run.

2.6 Working With CNC Machine Constraints

Every CNC machine has quirks, from tool size limits and axis travel to material hold-down and surface finish variations. These factors shape how parts are put together, design through fabrication.

Accounting for these factors early helps avoid costly rework, improves accuracy, simplifies setups, and ensures a smoother production process and fewer surprises.

CNC Machine	Constraints	Design Tips
3D Printing	Limited resolution, material shrinkage, and layer adhesion issues can affect print quality.	Use supports for overhangs and ensure proper bed adhesion to prevent warping.
CNC Router	Rounded internal corners, the need for dogbones or T-bones, and chatter from high speeds or poor clamping can impact CNC results.	Use dogbones or relief cuts to prevent splitting and ensure fit.
CNC Plasma	Poor tiny detail handling, post-processing for slag, and angled finishes on thick steel are common issues.	Avoid fine serif fonts or tiny cutouts.
CNC Waterjet	Cuts most materials but with edge taper, slower than lasers on thin stock, and abrasive wear needs attention.	Offset paths to maintain critical dimensions.
CNC Laser	Burns organics, needs caution for reflective surfaces, and can discolor light materials.	Use masking tape to reduce smoke marks.

A tightly nested 4 × 8 ft sheet of 3/4-inch MDF. Almost zero waste.

2.7 Prepping Files for CNC Success

Before you even think about making that first cut, everything happens in the digital world. Whether you're working in 3D CAD or vector programs like Adobe Illustrator, CorelDRAW, or Inkscape, this stage is where the magic, or misery, begins.

Just because your design looks slick on your monitor doesn't mean your CNC machine will get it.

It's like crafting a gourmet meal and serving it to a toddler, with a wrench instead of a fork. Think layers, operations, and depths. What's being cut? What's being engraved? What should stay untouched?

Machines don't think, they just follow orders. Bad input = expensive output. They require neat, clean 2D vector files (hello, EPS!), even if you started in 3D. Trust me, a well-prepared file keeps you sane when it's 2 am, you're still cutting, and the customer arrives in 5 hours for pickup.

For years at CNCKing.com, my workflow was a wild ride, designing in Autodesk 3DS Max, exporting to Illustrator, then VCarve for my ShopBot Desktop where I'd do further CNC router specific editing. Nothing was parametric.

It was like a game of digital hot potato, but it worked!

Good File Practices

Avoid overlapping lines: Especially for cut paths, they confuse the machine and ruin your parts.
Use closed paths: Open shapes lead to incomplete or messy cuts.
Color-code your layers: For example, red = cut, black = engrave. Keep it consistent.
Name layers clearly: Include dimensions in both millimeters and inches for universal understanding.
Attach a PDF copy: Cross-reference metric and imperial measurements so nothing gets lost in translation.

Be kind to your future self, or your machine operator. Clear labeling now saves major headaches later. Mistakes come from assumptions, avoid them as much as possible.

Proper nesting = scrap control mastery.

⬚ 2.8 Success Story
Custom Floor Vents

Mixing materials? Clamp hard, glue harder.

In this case study, I had the pleasure of creating custom floor vents that were a perfect blend of woodworking and metal fabrication, because why settle for boring vents when you can have functional art?

The goal was simple: make vents from solid wood with a stainless steel backing that wouldn't warp and still had perfectly aligned vent openings. Easy, right?

First, I prepped the materials like a pro. Then, it was time to CNC route the wood, followed by CNC plasma cutting the steel backing, because why not use two machines to do the job of one?

The real magic came in the gluing stage, where I meticulously combined the two materials like a mad scientist with a penchant for perfection.

The result? A floor vent so precise it might be the fanciest feature in your entire house, sorry, floors.

Keeping it flat before and after glue? Easier said than done.

Process Breakdown

1. Material Selection and Preparation

I used walnut-like wood, removed a layer to prevent warping, and added a stainless steel backing for extra muscle and long-term durability. Material prep wasn't just about aesthetics, it was about structural integrity.

2. CNC Routing and Plasma Cutting

The CNC router carved crisp slats, while plasma cutting delivered a stainless steel backing ready to take whatever abuse your floor dishes out.

3. Challenges and Solutions

Warping During Plasma Cutting:
I mitigated warping caused by heat expansion and contraction by using a thicker gauge (material thickness) than was initially envisioned for this project. Adjusting the cut order and allowing for cooling between passes also helped maintain flatness.

Production vs. Prototypes:
Prototype testing revealed issues like excessive glue use and plasma cutting order. Adjustments were made to both during production.

Clamp Shortage:
Running out of clamps caused a delay, resulting in the production being split into two batches. You can never have too many clamps in a shop!

4. Final Assembly and Quality Control

I bonded the stainless steel to the wooden frame for a flawless fit. The result? A vent so smooth it elevates the whole room. Though, you'd never see the stainless backing.

These custom floor vents were unique in every way, design, material, and the mad dash to finish them before the house reno slammed shut.

It felt less like manufacturing and more like a relay race against the final city inspector, clipboards in hand and no mercy in sight!

CNC plasma + CNC router = custom steel-backed oak floor vents.

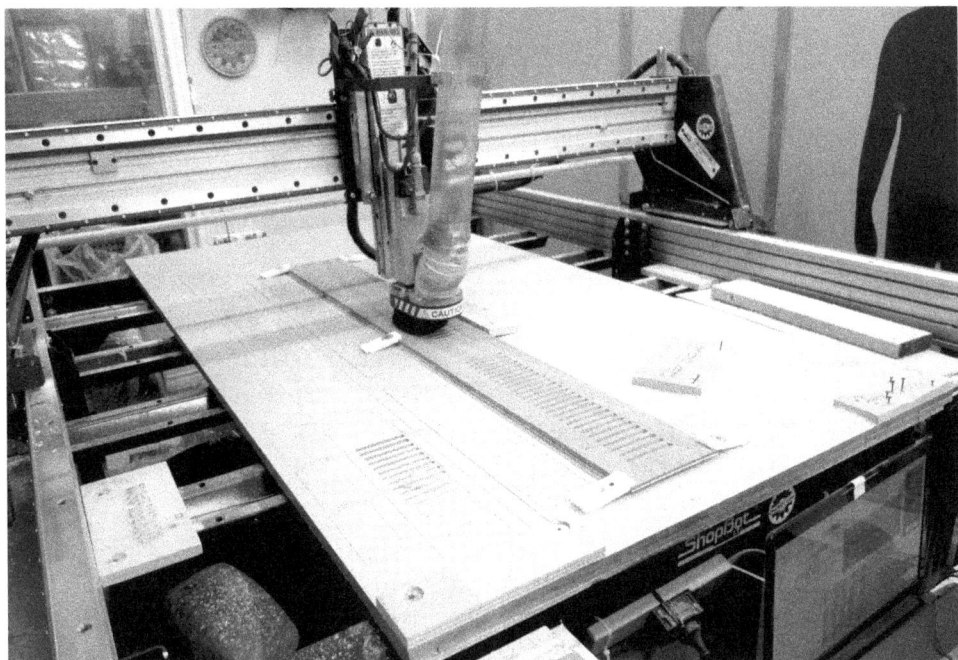

Anything full of holes is basically a structural dare.

Things I Had to Learn the Hard Way

Precision in Plasma Cutting

I learned the importance of careful cut sequencing to prevent warping when cutting metal with tons of small holes. Heat is your #1 enemy, especially with stainless.

Prototype Testing

Prototypes revealed production pitfalls, allowing improvements for the final version.

Clamps and Holding Systems

The project emphasized the importance of having enough clamps to secure materials during the drying process.

Another One Off the Table

This project reinforced how the right tools, materials, and prototyped techniques can turn even everyday details into standout features. Was a fun project.

🏅 2.9 Practical Takeaways from the Shop

Always run a tolerance test: Machines, materials, and even the weather can sabotage your project. It's like CNC insurance.

Think like the assembler: (Hint: it's probably you).

Plasma cutters and CNC routers can't produce sharp internal corners: So expect extra finishing work, it's smarter to design around real-world outcomes, not just what looks perfect in your CAD render.

Label your parts, prep clean cut files, and always account for kerf: Few things slow down assembly more than mystery parts, mismatched slots, or tabs that snap under pressure.

❌ 2.10 Chapter 2 Quiz

Time to prove what you know, or just circle C and cross your fingers?

1. What's the main focus of "Design for Manufacturing"?
    A) Aesthetics of a final product
    B) Making parts easier to prototype
    C) Optimizing a design for the fabrication process
    D) Choosing the most expensive materials

2. What's kerf?
    A) A type of wood joinery
    B) The error margin in file prep
    C) The material removed by the cutting tool
    D) The radius of a toolpath

3. Which tool has the narrowest kerf?
    A) CNC router
    B) CNC plasma
    C) Waterjet
    D) Laser

4. What if kerf is ignored in tab-and-slot design?
    A) The tab floats in mid-air.
    B) The tab and slot won't fit properly.
    C) The tab will disappear.
    D) The laser will stop mid-cut.

5. Which tool has the widest kerf in thick stock?
    A) Laser
    B) Plasma
    C) Waterjet
    D) Shear

6. Which design mistake do beginners often make?
    A) Accounting for tool diameter
    B) Designing for the end result only
    C) Using masking tape on plywood
    D) Using a PDF instead of a DXF file

7. What's key to good nesting?
    A) Leaving large spaces between parts
    B) Arranging parts for easy removal
    C) Tight packing with minimal gaps
    D) Using a single material type

8. Which statement is true about bad nesting?
    A) It reduces production cost.
    B) It optimizes tool change paths.
    C) It wastes material unnecessarily.
    D) It improves speed of production.

9. What's an interference (or friction) fit?
    A) A fit where parts are loose
    B) A fit requiring force to assemble
    C) A fit with no tolerances
    D) A misaligned toolpath

10. In CNC routing, why are dogbones or T-bones used?
    A) To fix machine vibration
    B) To speed up toolpaths
    C) To deal with rounded internal corners
    D) To reduce kerf

11. What's the purpose of offset paths?
    A) To ignore toolpath complexity
    B) To simulate aesthetic design in CAD
    C) To compensate for kerf or tool diameter
    D) To bypass post-processing

12. What's a typical issue with plasma cutting?
    A) Over-polished surfaces
    B) Heat warping
    C) Toolpath misalignment
    D) Rounded corners

13. Which file format is commonly required by CNCs?
    A) .JPG
    B) .MP4
    C) .EPS
    D) .DOCX

14. What's a best practice for organizing CNC design files?
    A) Using layers and clear labels
    B) Embedding 3D renders
    C) Drawing in RGB mode
    D) Using only imperial units

15. What does the acronym CAD stand for?
    A) Computer Aided Drafting
    B) Creative Art Design
    C) Computer Aided Design
    D) Conceptual Assembly Drawing

16. Which CNC tool is most prone to burning organics?
    A) Waterjet
    B) CNC plasma
    C) CNC router
    D) Laser

17. What do overlapping lines cause in CNCs?
    A) Faster cuts
    B) Enhanced resolution
    C) Incorrect tool movements or double cuts
    D) More accurate part fit

18. What should you do before running a full CNC job?
    A) Order new software
    B) Prototype and test tolerances
    C) Skip setup and let the operator decide
    D) Use grayscale color layers

19. Why is prototyping different from DFM?
    A) Prototyping ignores real-world constraints.
    B) DFM is more creative.
    C) Prototyping tests the concept before production.
    D) Prototyping replaces CAD.

20. Why was stainless steel added to the wooden vent?
    A) For aesthetics
    B) To act as a handle
    C) To prevent wood warping and provide support
    D) To reflect heat

CNC lasers turn plain Pine into pure magic.

✘ 2.11 Answer Key + Explanations

Check your answers here, let's see where you were on track or off-axis.

1C – *Design for Manufacturing is about making things that are not only creative but manufacturable with the machines and materials you have.*

2C – *Kerf is the width of material removed during cutting, critical for part fit.*

3D – *Lasers have the narrowest kerf, often around 0.1 mm, making them ideal for detailed precision work.*

4B – *Without kerf compensation, you'll either get too tight or too loose fits.*

5B – *Plasma tools generally remove more material due to arc width (1–2 mm or approximately 3/64–5/64 inch).*

6B – *Newbies often forget to consider how parts are assembled or cut.*

7C – *Efficient nesting maximizes material use and reduces waste.*

8C – *Bad nesting leaves unused space that adds cost.*

9B – *This is typical in press-fit or permanent assembly designs.*

10C – *Router bits can't make sharp inside corners, so dogbones (and other methods) provide clearance.*

11C – *Offsetting paths ensures accurate part dimensions despite tool size.*

12B – *Plasma's high heat can warp thin or small metal pieces.*

13C – *This vector format is widely accepted by most CNC machines.*

14A – *This ensures the CNC operator knows what to cut, engrave, or ignore.*

15C – *CAD is the foundation of digital part modeling before CAM or CNC.*

16D – *Lasers can burn wood and leather, masking can help reduce this.*

17C – *Overlapping lines confuse the machine, leading to potential errors.*

18B – *Testing your design ensures accurate fit and identifies problems early.*

19C – *DFM is for production; prototyping is for testing and refining ideas.*

20C – *Stainless backing stabilized the wood in CNCROi.com's custom floor vent.*

Masking, because no one likes sanding.

Glass plus laser? Classy, as long as it's masked first.

Channel system under my ShopBot spoilboard for vacuum.

Letting the sacrificial board cure, under pressure!

Chapter 3
Smart Material Selection
Balance Budget, Strength, and Surface Appeal

Smart material selection is where great projects begin, and where bad ones quietly fall apart. Balancing cost, strength, and finish isn't optional, it's the foundation of fabrication that works.

Not all plywood is equal, and neither is stainless steel. Some materials are heroes; others waste time, tools, or both. This chapter dives into choosing the right one, juggling price, durability, aesthetics, and machinability like a circus act with no safety net. Choose wisely. The project's fate is sealed long before the first cut.

🗩 *Use the wrong material and your project becomes a very expensive lesson in disappointment.*

Made 500 coasters, shipped 400. 100 became branded firewood.

3.1 Why Does Material Selection Matter?

Materials are the foundation of everything you build, get it wrong, and things warp, snap, or melt. Get it right, and you look like a god.

Before you cut, weld, or glue, know how your material behaves, these key properties will save you time, tools, trouble, and money.

Term	Meaning	Why It's Important
Elasticity	Ability of a material to return to its original shape after deformation	Affects material's performance under stress and load
Hardness	Resistance of a material to indentation or scratching	Impacts wear resistance, tool choice, and longevity
Tensile Strength	Maximum stress a material can withstand while being stretched or pulled	Critical for structural integrity and durability
Gauge (Metal)	Thickness of metal sheet	Critical for strength and compatibility
Grain Direction	Orientation of fibers in wood, rolling direction or grain flow in metal	Affects strength, warping, and appearance
Machinability	How easy it is to cut or shape	Impacts time, wear on tools, and cost
Moisture Content	Amount of water in a material (mostly wood)	Can warp or split material post-cut
Thermal Conductivity	How well heat moves through a material	Impacts welding, laser settings, and thermal bonding outcomes

👎 *Choose Material for Looks, Not Performance.*

🖥 3.2 Dealing with Large Suppliers

For small businesses, sourcing materials from large suppliers can be challenging, especially for smaller quantities.

Here are some strategies to help secure the best deals:

*1. **Be Prepared:** Know your material needs, pricing, and check for minimum order quantities (MOQs).*

*2. **Leverage Small Business Benefits:** Emphasize your speed and flexibility. "I can work with your offcuts, just don't call me a "scrap artist"!*

*3. **Build Relationships:** Foster rapport with reps for better deals and customer support.*

*4. **Negotiate Discounts:** Ask about bulk pricing or flexible payment terms, even for small orders.*

*5. **Seek Alternatives:** Consider local suppliers or group purchases to meet MOQs.*

*6. **Know When to Walk Away:** Don't settle. If the terms stink, walk, there's always another supplier who wants your business.*

*7. **Time Your Orders:** Order when suppliers are trying to meet sales goals for potential discounts, generally at the end of the month.*

*8. **Explore Niche Suppliers:** Look for businesses focused on small batches and better customer service.*

*Play your cards right, and even the big suppliers will start offering you volume discounts, maybe even a few "lost" pallets if you ask nicely. A good supplier will find a way to help you when timelines get tight, inventories run dry, or you suddenly need triple the material you thought you did. **Relationships matter.***

3.3 Material Categories

Design and fabrication are not just about appearance. They are about durability, performance, and how long something is meant to last. Smart material choice = fewer headaches, cleaner cuts, and a smoother build.

Quite often, the customer will come to you with a material preference and once you delve deeper, you may realize there is another that's more suitable to the project at hand.

Category	Examples	Common Traits	Typical Uses
Composites	Carbon fiber, Fiberglass	High strength-to-weight, layered construction	Aerospace, sports equipment, tooling
Foams	EVA, Polystyrene, Urethane	Lightweight, shock-absorbing, easy to shape	Packaging, props, insulation
Metals	Steel, Aluminum, Brass	Strong, conductive, recyclable	Structural parts, enclosures, brackets
Paper-based	Cardboard, Chipboard, Kraft	Inexpensive, lightweight, easy to cut	Packaging, stencils, mockups
Plastics	Acrylic, ABS, Polycarbonate	Lightweight, insulative, machinable	Signage, prototypes, covers
Rubbers / Elastomers	Silicone, Neoprene, EPDM	Flexible, resistant to wear and chemicals	Gaskets, seals, vibration dampening
Woods	Oak, Plywood, MDF	Aesthetic, directional grain, moisture-sensitive	Furniture, cabinetry, decorative panels

3.4 Wood & Wood Products

Wood: the original all-rounder in fabrication. But not all wood is equal, some's strong, some's pretty, and most eventually warp like a potato chip.

Type	Strengths	Limitations
Hardwood	Beautiful, durable	More expensive, can move over time
MDF	Smooth, cheap, great for engraving	Dusty, weak edges, hates moisture
Plywood	Strong, consistent, great for routing and laser	Can delaminate or warp if cheap
Softwood	Lightweight, cheaper	Not as strong or stable

⚠️ Look for cabinet-grade, not "construction" ply.

This is how I feel every day I'm able to work in my shop!

3.5 Plywood ≠ Plywood

That $30 sheet of pine plywood from a big-box store? Probably warped and full of voids. Your CNC will find every imperfection and create new ones just to mess with you. Stick with higher-grade material from trusted suppliers.

Grade	Description	Use Case
CDX	*Low-quality plywood with visible defects, often used for exterior sheathing.*	*Cheapest $ Structural use, outdoor applications*
C	*Plywood with minor imperfections and some visible knots or patches.*	*Low $$ General construction and utility*
B	*Higher quality, fewer visible knots, some filler. Suitable for visible surfaces.*	*Mid $$$ Cabinetry, furniture, and indoor use*
A	*High-quality plywood with smooth, blemish-free surface and minimal imperfections.*	*Expensive $$$$ High-end furniture, fine cabinetry, visible surfaces*
Marine	*Premium plywood treated for high moisture resistance, typically made from durable woods.*	*Luxury $$$$$ Boat building, high-moisture environments*

3.6 Routers Rule for Thick Plywood Cuts

CNC routers love thick, glue-heavy plywood, no burn marks, just clean cuts. They're faster on dense stuff, perfect for layered designs like metal inlays, and chew through cheap, replaceable bits. Plus, they kick up dust, not fumes.

Lasers are great for detail, but routers win on strength, edge quality, and versatility, your plywood MVP... but of course lasers can still engrave the blanks made with a router! Remember, CNCs don't just cut, they collab!

3.7 Plastic Products

Not all plastics are flimsy, some, like carbon-fiber and fiberglass, could probably win an arm-wrestling match against your toolbox.

Knowing what your material can (and can't) do is key to picking the right one for a project.

Type	Strengths	Limitations
ABS	Tough, impact-resistant, easy to process	Not UV resistant, can warp under heat
Acrylic (PMMA)	Beautiful edges, engraves well, comes in colors	Brittle, can crack under stress
Carbon Fiber	Extremely strong, lightweight, rigid	Brittle, expensive, difficult to cut (may need diamond bit)
Delrin (Acetal)	Strong, low friction, highly durable, easy to machine	Expensive, can be difficult to glue
Fiberglass	Lightweight, strong, good chemical resistance	Difficult to machine, hazardous dust
HDPE / UHMW	Slippery, durable, easy to route	Poor adhesion, can warp
Nylon	Tough, abrasion-resistant, flexible	Absorbs moisture, can be hard to machine
Polycarbonate	Super strong, transparent	Expensive, hard to laser cleanly
Polypropylene (PP)	Lightweight, chemical-resistant, easy to weld	Low strength, poor surface finish for painting
PVC	Cheap, cuts well	Toxic fumes when laser cut, avoid!

3.8 Metal Products

Metals are the muscle of any project, picking the right one is like choosing the right superhero (more on that later).

Some fight corrosion, some behave better than a Shih Tzu at obedience school, and some just look awesome. Strength, cost, weight, it's all about matching the metal to the mission. All metals can be welded. Some simply require greater skill in both pre weld and post weld processes.

Type	Strengths	Limitations
304 Stainless Steel	Corrosion-resistant, excellent for most applications	Prone to stress corrosion cracking, slightly magnetic
316 Stainless Steel	Superior corrosion resistance (e.g., marine environments)	More expensive than 304, slightly harder to machine
Aluminum	Lightweight, corrosion-resistant	Gummy when routed, hard to weld without care
Cast Iron	Excellent machinability, durable, vibration-damping	Brittle, prone to cracking, heavy
Copper / Brass	Aesthetic, good conductivity	Costly, reflect laser light (use fiber)
Nickel Alloys	Excellent corrosion resistance, heat-resistant	Expensive, difficult to machine
Steel (Mild)	Strong, welds well, inexpensive	Heavy, rusts easily without coating
Stainless Steel (others)	Beautiful, corrosion-proof	Expensive, harder to cut and weld
Titanium	Extremely strong, corrosion-resistant, lightweight	Expensive, difficult to machine, requires special tooling
Tool Steel	Hard, wear-resistant, holds sharp edges	Difficult to machine, expensive

3.9 Choosing Materials by Application

Choosing the right material depends entirely on the application.

For outdoor signs, avoid MDF, it doesn't hold up to moisture, even if you seal it. Instead, use stainless steel or solid surface materials like Corian for durability. Food-safe products require hygienic, easy-to-clean options like HDPE or stainless steel. Laser etching best for stainless, router carving for HDPE.

Furniture benefits from marine-grade plywood or hardwood for strength and longevity. Art displays are driven by the artist's vision, so materials vary widely.

For industrial parts, material selection often depends on budget, performance, and turnaround time.

Each use case has its own demands, so matching material properties to the application is key to long-term success.

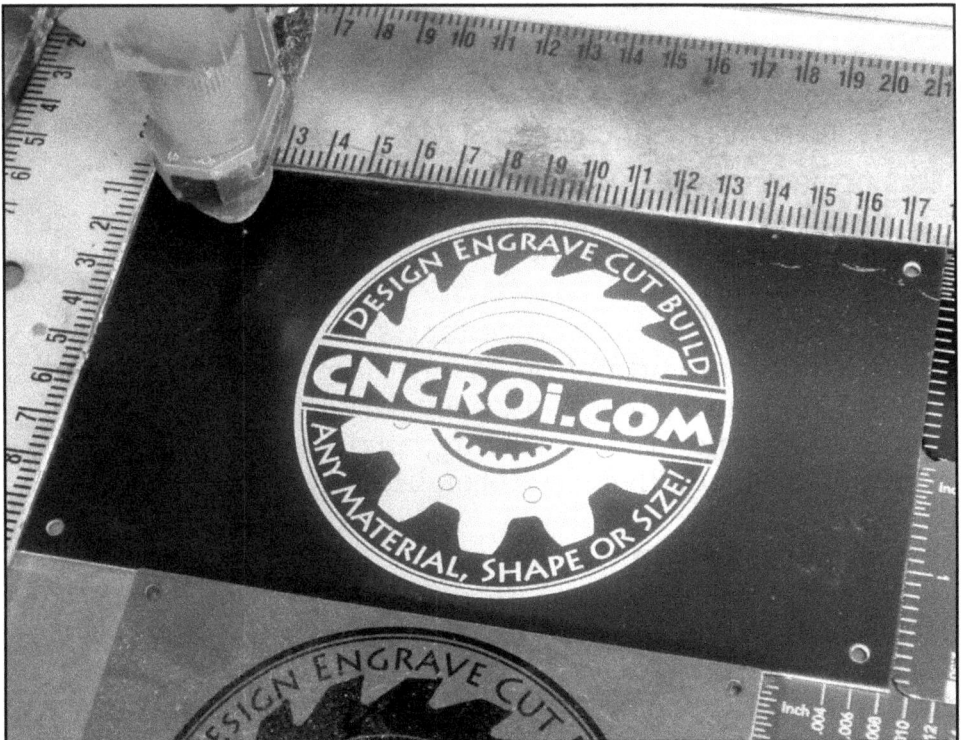

CO_2 + anodized aluminum = the tuxedo of laser engraving.

Bone Corian Outdoor Plaques

CNC-routed at 1" (25.4 mm) thickness; laser-engraved to 1/8" (3.175 mm) depth and filled with black enamel paint.

This project was all about creating custom bone Corian plaques for outdoor display, because nothing says "tough" like a material called "bone".

Corian was the perfect choice for its outdoor durability, so these plaques wouldn't fade faster than your favorite pair of jeans. The result? Five plaques, each crafted with a mix of CNC routing and laser engraving, all designed to look stunning and weather the storm, literally.

Process Breakdown

1. Material Selection and Preparation

Corian, a solid surface material, was the perfect mix of strength, easy maintenance, and holding fine details like a pro. The plaques were made from one-inch thick Corian.

Prototyping: because guessing only looks good after Photoshop.

Since I like to keep things interesting, I bonded two half-inch slabs together for extra durability. The laser engraving cut deep, leaving long-lasting details even if the paint fills eventually decided to take a vacation.

2. CNC Routing and Laser Engraving

I kicked things off with the CNC router, cutting out the basic shapes, because who wants to carve things by hand anymore? After that, it was time for laser engraving, where the magic (and the smoke) happened, carving out the fine details with precision. It wasn't a quick process.

3. Challenges and Solutions

Old Photographs and Resolution:
The primary issue? Ancient, low-res photos. The laser engraver only speaks in black-and-white, so I had to give these old images a makeover, kind of like what Photoshop does to your ex's Facebook photos.

Test plates helped determine which images were engraving-worthy and which needed more sharpening.

Paint Filling:
Choosing the right paint was a drama. UV printing was off the table, too risky with the outdoor exposure. So, I went old school, using enamel paint for the engraved areas.It lasts longer than my last tooling strategy.

4. Final Assembly and Quality Control

The plaques came together beautifully, ready to face the elements without breaking a sweat. Outdoor exposure? Bring it on all day, every day.

These were built to handle sun, rain, snow, and maybe even the occasional rogue sprinkler.

These plaques will outlast your neighbor's garden gnome collection from the 1960s, and probably still look good when that gnome army has crumbled into yard dust.

Masking helped with paint... kinda like an umbrella in a hurricane.

Coated like they're starring in a paint commercial.

Finished Corian plaque, after hours of cleaning-up.

53

Things I Had to Learn the Hard Way

Pre-Test Images

Testing images on scrap before committing to the expensive Corian was an effective way to refine the designs and avoid costly mistakes.

Laser Limitations

Enhancing image resolution and contrast was crucial, since the laser can only interpret black-and-white images.

Corian Durability

Corian proved to be an excellent material for outdoor signage, offering both durability and aesthetic appeal.

Time Management

The laser engraving process was slow and required multiple passes to achieve the necessary depth, but the precision of the results made the time investment worthwhile.

Another One Off the Table

This custom bone Corian plaque project really proved that picking the right materials is half the battle, add in some precision tools like CNC routers and lasers, and you've got yourself a winner.

Combine thoughtful design with reliable equipment, and even tricky projects start to feel manageable, even under tight deadlines.

Sure, the old photos and lengthy engraving process tested my patience, but the end result was totally worth it, clean lines, deep cuts, and zero compromises.

The plaques now boast both beauty and durability, ready to face the elements and look good doing it, for years to come, no less! These weren't just built to last, they were built to impress.

🏅 3.11 Practical Takeaways from the Shop

14 gauge (ga) isn't the same as 10 gauge, and 10 gauge isn't the same as 10 mm: Always convert and double-check, or your design might end up with some unexpected gaps.

Beauty's great, but function comes first: If your metal can't handle the job, you'll end up with a pretty piece of junk.

Wood is like that unpredictable friend: Plan for material movement or risk dealing with warped pieces down the line.

Too much power when engraving plastic: Can turn your project into a melted mess. Lower your settings and always test first, your edges will thank you.

❌ 3.12 Chapter 3 Quiz

Take this materials quiz to see if you're a genius, or just a lucky guesser with a sharp eye for metal!

1. Why is material selection crucial in manufacturing?
    A) It keeps the product looking good.
    B) It affects how easily a material can be shaped.
    C) It impacts durability, machinability, and cost.
    D) It boosts environmental sustainability.

2. Which property restores shape after bending?
    A) Hardness
    B) Elasticity
    C) Grain Direction
    D) Tensile Strength

3. What's the importance of grain direction in wood?
    A) It determines the color of the wood.
    B) It determines how easily the wood can be painted.
    C) It indicates how the wood should be cut.
    D) It affects the strength, warping, and appearance.

4. What material is strong, light, and used in aerospace?
   A) Foams
   B) Composites
   C) Metals
   D) Plastics

5. Which metal resists corrosion best?
   A) 304 Stainless Steel
   B) Aluminum
   C) 316 Stainless Steel
   D) Cast Iron

6. What's the biggest downside of using MDF outdoors?
   A) It absorbs moisture like a towel and swells.
   B) It is too expensive.
   C) It is too heavy.
   D) It is not aesthetically appealing.

7. What's the best material for tough, chemical-resistant seals?
   A) Silicone
   B) Acrylic
   C) Steel
   D) Nylon

8. What does tensile strength measure in a material?
   A) Its resistance to indentation or scratching
   B) Max stress a material can take when stretched
   C) How easy it is to cut or shape
   D) The amount of water it can absorb before warping

9. What's a common trait of aluminum and steel?
   A) They are insulative.
   B) They are biodegradable.
   C) They are easy to engrave.
   D) They're lightweight and resist corrosion.

10. Why does metal gauge matter?
   A) It determines the material's color.
   B) It indicates thickness and part fit.
   C) It indicates how shiny the metal will be.
   D) It measures the weight of the material.

11. Which material is best for food-safe products?
   A) HDPE or Stainless Steel
   B) MDF
   C) Acrylic
   D) Polycarbonate

12. What's a major drawback of using ABS plastic?
   A) It is not UV resistant and can warp under heat.
   B) It is brittle and cracks easily.
   C) It is too expensive to use.
   D) It cannot be shaped easily.

13. Which one is NOT an attribute of stainless steel?
   A) Corrosion-resistant
   B) Expensive
   C) Sometimes magnetic
   D) Lightweight

14. Why test on scrap before using expensive material?
   A) To speed up the fabrication process
   B) To avoid mistakes and refine designs
   C) To test the material's paint adherence
   D) To reduce laser cutting time

15. What material made up the Bone Corian plaques?
   A) MDF
   B) Acrylic
   C) Corian
   D) Aluminum

16. What's the biggest challenge in the Bone Corian project?
   A) Working with old, low-resolution photographs
   B) Cutting the Corian efficiently
   C) Applying paint that would not wear away
   D) Matching the plaque size to design specifications

17. What's correct about the case study's CNC routing?
   A) It is used to apply paint to the material.
   B) It is the first step to cut the plaques' shape.
   C) It is used to create the final image engraving.
   D) It requires high power settings to work efficiently.

18. Why avoid construction plywood for fine work?
    A) It is too expensive.
    B) It is prone to warping and contains imperfections.
    C) It is too strong for certain applications.
    D) It is difficult to cut with CNC machines.

19. Why choose high-grade plywood over low-grade?
    A) It is cheaper.
    B) It is more durable than other materials.
    C) It is easier to laser engrave.
    D) It provides consistent quality and fewer defects.

20. How do you prevent warping or cracking in wood?
    A) Use a thicker gauge.
    B) Always pre-test the material.
    C) Consider moisture content & grain direction.
    D) Increase cutting speed.

These pine fish look great, just don't try to eat them.

✖ 3.13 Answer Key + Explanations

Let's see if your choices held up, like good materials should.

1C – *Material choice affects performance, cost, and lifespan.*
2B – Elasticity is a material's ability to snap back after stress.
3D – *Wood grain direction impacts strength, machining, and moisture response.*
4B – *Composites like carbon fiber offer high strength-to-weight ratios, ideal for aerospace.*
5C – *316 stainless resists corrosion, especially in marine settings.*
6A – *MDF absorbs moisture, making it poor for outdoor use.*
7A – *Silicone is flexible and chemical-resistant, great for seals and gaskets.*
8B – *Tensile strength is a material's resistance to pulling forces.*
9D – *Aluminum and steel offer strength, low weight, and corrosion resistance.*
10B – *Gauge measures metal thickness, affecting strength and fit.*
11A – *HDPE and stainless are safe, durable, and easy to clean for food use.*
12A – *ABS isn't UV-resistant and warps with heat, unsuitable outdoors.*
13C – *Stainless is usually non-magnetic; 304 may show slight magnetism.*
14B – *Testing designs on cheap materials saves cost before using premium ones.*
15C – *Corian was used for its durability and fine engraving capability.*
16A – *Laser engraving required enhancing low-res photos for clarity.*
17B – *CNC routing shaped plaques before engraving.*
18B – *Low-grade plywood warps and has defects, bad for precision cuts.*
19D – *High-grade plywood cuts cleanly and consistently.*
20C – *Managing wood grain and moisture prevents warping and cracking.*

This chip's grain is stronger than my poker face.

Guaranteed to raise stakes... or eyebrows.

Chapter 4
CNC Routing That Shapes Ideas
Cutting Paths, Crafting Possibilities

CNC routing shapes more than material, it shapes ideas. From signs and furniture to enclosures and art, CNC routers turn digital designs into real-world results by cutting paths and unlocking possibilities.

They're custom manufacturing workhorses, versatile, scalable, and ideal for flat or lightly contoured materials.

Let's demystify CNC routing, from generating toolpaths to choosing bits and avoiding rookie mistakes like chipping, blowout, or even fire. We'll cover 2D profiling, pocketing, 3D carving, and basic fixturing for clean, accurate results. Knowing where we came from makes it clear why CNC is the gold standard for modern production.

CNC router + giant mold = instant gym membership.

💬 *Think of your CNC as a loyal dog, it'll fetch the stick, even if it's on fire.*

4.1 What's a CNC Router?

A CNC router is a computer-controlled wood chipper that moves a spinning knife along multiple axes, typically X (left/right), Y (forward/backward), and Z (up/down).

It's ideal for working with sheet goods like wood, plastics, and softer metals like aluminium. Think of it as a hyper-precise robot sculptor... except less romantic and more dusty. CNC mills are typically designed for machining metals and harder plastics. Same idea, different tuning.

Since CNC routers were basically the "first CNC," people often say just "CNC" when referring to them, not that 6-axis robot in your shop cutting titanium and welding it together. Technically, your office printer is also a CNC, just not as fun to watch.

👎 *Not Letting the Design Lead the Tool.*

4.2 Why Does CNC Win (Most of the Time)?

While manual routing still has value for quick fixes or artistic work, CNC routers elevate the craft. They turn years of skill into minutes of programming, delivering consistent, repeatable precision, without sacrificing creativity.

Feature	Manual Routing	CNC Routing
Accuracy	Depends on the operator	Computer-controlled precision
Repeatability	Hard to match cuts	Identical results every time
Complexity	Limited by skill	2D, 2.5D, and full 3D possible
Speed	Slower, more setup time	Faster, more consistent
Skill Ceiling	Steep learning curve	Learn CAD/CAM and you're set

4.3 Router vs. Spindle

For real production, ditch the noisy router and get a spindle, your ears (and your parts) will thank you.

Router: Loud, cheap, and needs lots of naps.

Spindle: Quiet, precise, and built to hustle nonstop.

4.4 What Can CNC Routing Do for YOU?

CNC routers tackle 2D cuts, pockets, drilling, 3D contours, and precision V-carving, perfect for making signs, joinery, decorative carvings, custom sculptures, and parts that actually fit the first time.

Whether you're cutting sheet goods, carving hardwood, or prototyping with foam, CNC routers turn your digital designs into real-world results with speed and repeatability.

Operation	What It Means & Why It's Useful	Design Considerations
2D Profiling	Cuts out shapes from flat sheets. Ideal for signs and panels.	Use tabs or onion skins to hold parts in place during cutting.
3D Contouring	Carves complex curves, great for sculptural or mold-making work.	Requires a 3D model, ball-nose bits, and allows for longer run times.
Drilling	Creates vertical holes for mounting, alignment, or joinery.	Use peck drilling for depth; match drill size to fastener type.
Pocketing	Clears flat-bottomed areas for inlays, trays, or joints.	Choose bit size and stepover to reduce tool marks or chatter.
V-Carving	Variable-depth line carving for detailed text and designs.	Needs a V-bit; sensitive to Z-height; achieves sharp internal corners.

Unless you're doing lots of cabinetry, you can just use a normal bit for drilling operations until you have enough volume to justify a drill unit or drill bank (lots of different drills). It's a little slower and tougher on the spindle, but perfectly fine for limited use.

Keep in mind that a CNC router isn't just great at cutting and carving, it's also your shop's unsung hero for flattening wood.

Whether you've got a twisted live-edge slab that looks like it survived a small tornado or a laminated mold blank with more bumps than a dirt road, the router's got you.

Need to surface a tabletop that's doing the wave? Router. Working with hardwood that thinks it's still a tree? Router. It's like having a digital lumberjack with OCD for flatness.

Just slap on a surfacing bit, zero it out, and let the router do its thing, methodically shaving it dead-flat and parallel.

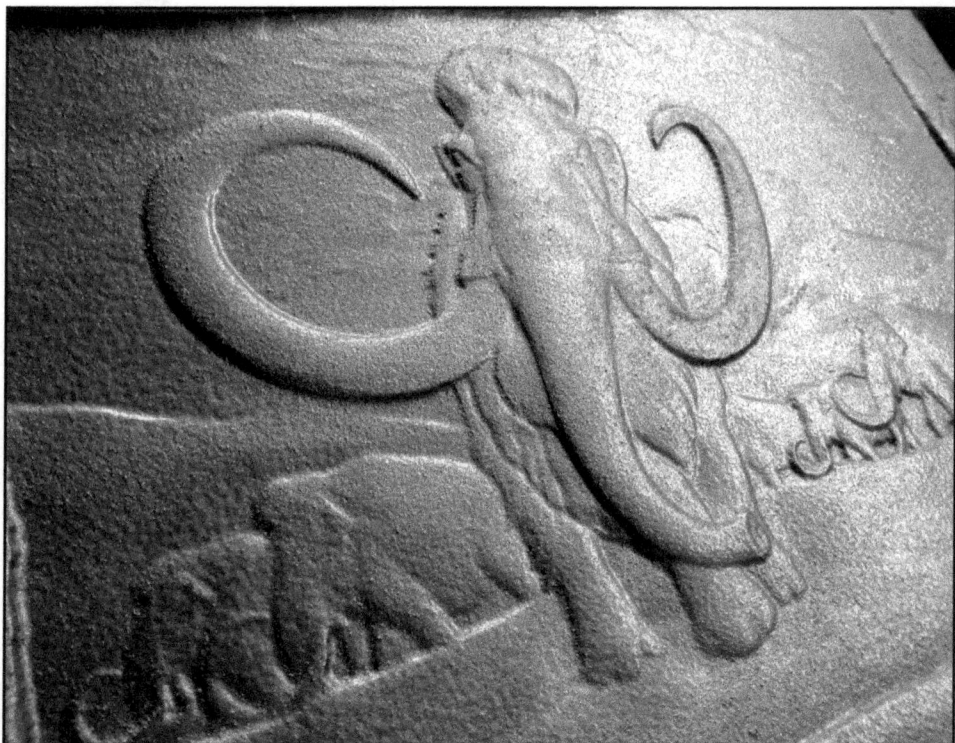

Tiny router, giant mammoth, MDF never stood a chance.

4.5 Tooling 101: Bits & End Mills

You wouldn't use a chainsaw to carve a spoon (unless you were showing off). Same goes for CNC bits, tool choice is crucial. Tool diameter affects speed, depth, and precision.

Material	Tool / Use	Why Preferred / Avoid
Aluminum	Carbide or coated HSS end mill, smooth cuts	Sharp tools reduce burrs; dull ones cause wear and rough edges.
Acrylic	O-flute or polished end mill, clean finish	Prevents melting; dull bits lead to rough edges and heat buildup.
Brass	Polished carbide or HSS, clean surfaces	Reduces friction; dull tools degrade finish and clog.
Copper	Carbide or HSS, low-friction smooth cuts	Carbide resists buildup; rough tools leave poor finish.
Foam	Drag knife or low-RPM router, tear-free cuts	Low speeds prevent tearing; high RPM causes roughness.
MDF	Downcut or compression bit, clean tops	Reduces splintering; upcut bits fuzz fibers.
Polycarbonate	Polished or upcut bit, crack-free cuts	Smooth results; dull tools overheat and crack material.
Plywood	Spiral upcut or compression bit, clean edges	Minimizes splinters; straight/downcut bits cause tear-out.
Stainless Steel	Plasma, fiber laser or waterjet, precise cuts	Heat distortion unless you are using flood coolant with bits.
Titanium	Coated carbide (TiAlN/TiCN), for tough cuts	Resists wear; low-grade HSS fails quickly.

⚠ Assume bits are dull, as they usually are.

4.6 Feed Rates & Speeds: Simplified

Keep in mind that you don't need to memorize formulas, but understanding the relationship between feed rates, RPM, and cutting depth is key to getting the best results. **Always test on scrap before full production!**

Key Terms Simplified:
- Feed Rate: How fast the bit moves, like pushing a shopping cart, slow or sprint.
- RPM: How fast the bit spins, like blender speed.
- DOC (Depth of Cut): How deep the bit cuts, like peeling a potato instead of coring it in one go.

Start conservatively and experiment to find the sweet spot.
- Faster feed + slower RPM = better chip evacuation on softer materials.
- Slower feed + higher RPM = cleaner finish on acrylics and hardwoods.
- Shallow passes = less tool wear, especially in dense materials.
- Deep cuts = faster production, but more strain on machines and bits.

⬇ 4.7 Chip Load ≠ Fancy Nacho

You want something more technical?

Chip load = the amount of material removed per tooth per revolution.
Too low? You burn the wood.
Too high? You break your bit.
Yes, it's math, but it's worth learning if you don't have time to play around.

⚠ *See Appendix D for a broader explanation.*

4.8 Workholding: Keeping Material in Place

Vacuum tables are the gold standard on high-end setups, but honestly, even basic screws, my trusty go-to for years, get the job done just fine. It's not always about fancy gear, it's about knowing how to use what you've got.

Method	Strengths	Limitations
Screws	Super solid	Leaves holes.
Double-Sided Tape	Clean top	Can shift, leaves residue.
T-Track and Clamps	Adjustable, reusable	Watch for bit collisions.
Vacuum Table	Fast and clean	Expensive, needs flat material.
Tabs	Bridges material	Must trim post-cut.

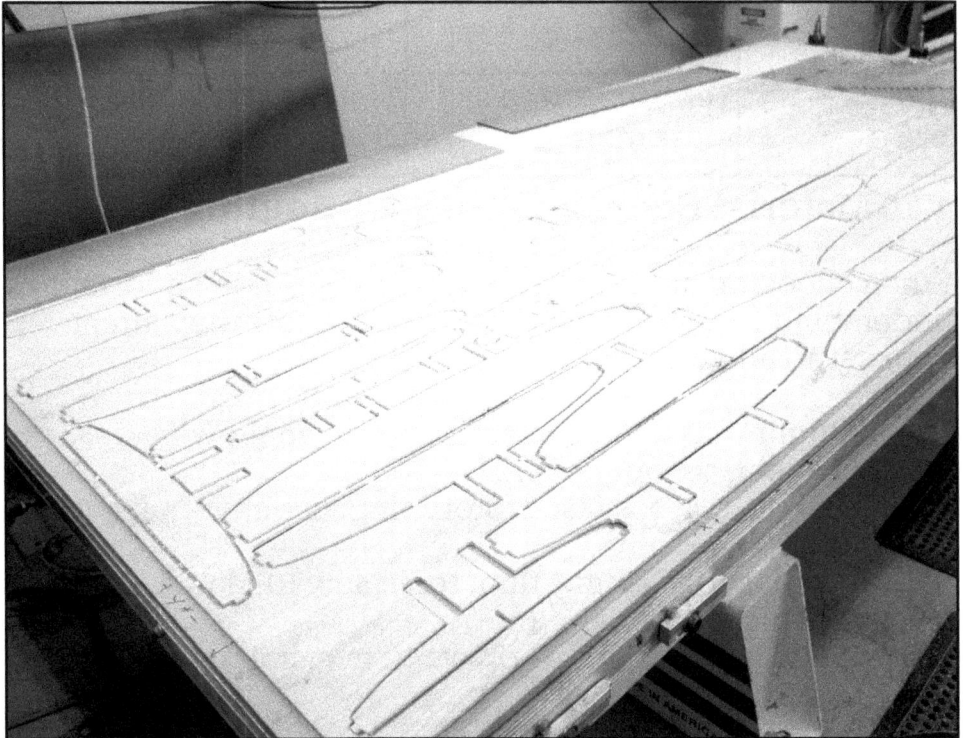

Vacuum table: the silent hero of floppy plywood.

4.9 Common Problems & Fixes

My fantastic little ShopBot Desktop can produce edge quality just as clean as my awesome, room-sized Thermwood, which also has an exponentially higher BMI. The difference? One fits on a bench. The other needs its own zip code.

A massive CNC just lets you mess up faster, is far louder, and doubles as a shop heater.

Problem	Likely Cause	Fix
Tear-out	Wrong bit direction or dull tool	Use downcut or compression bits.
Charring	Bit spinning too fast or too slow feed	Increase feed or decrease RPM.
Material Shift	Weak hold-down	Add clamps or better tape.
Bit Breakage	Cutting too deep or too fast	Use shallower passes.
Ridges in Cut	Tool deflection or loose gantry	Tighten everything & check overlap.

4.10 Material Considerations for Routing

When choosing materials for CNC routing, it is important to understand their unique properties and how they affect your setup and cutting process.

- **Aluminum:** Needs lube, slow feed, shallow passes, challenging but doable.
- **Foam:** Great for fast prototyping; use sharp bits, low RPM.
- **Plastics:** Use single-flute tools, low RPM to avoid melting.
- **Wood:** Choose stable plywood over solid wood for consistency.

⚠ See Q14 in the Q&A section for CNC routing metal.

4.11 Finishing Impacts Strategy

Plan ahead, your future self will thank you.

For clean edges, use a compression bit or apply a finishing pass. I prefer the finishing pass method.

Painting later? Don't stress over tabs and sanding; paint doubles as camouflage.

Joining parts? Keep tabs minimal and cuts tight unless you love "close enough" builds that involve hours of chisel time.

Visible edges? Watch your angular flare and router spin, nothing ruins a masterpiece like a crooked burn.

4.12 Toolpath Strategy Basics

Toolpaths are the route your machine follows to cut a part, and choosing the right one is key to boosting productivity. The wrong path increases wear, wastes time, and leaves ugly edges.

Strategy	When to Use It	Why It Matters
Climb Milling	CNC routers, mills	Produces a cleaner finish, less wear, more aggressive cuts.
Conventional Milling	CNC routers (for brittle materials)	Reduces chipping for delicate edges.
Lead-ins / Lead-outs	All CNC tools	Prevents burning or tool marks on start/stop points.
Onion Skinning	CNC routers	Keeps part from shifting; finish final pass last.
Ramping Cuts	CNC routers	Reduces tool wear by easing into cuts.
Tabbing	All cutting processes	Holds small or detailed parts in place.

⬇ 4.13 Tool Life Factors

Want your tool to live long and prosper?

- *Feed rate too low: Creates friction, heats up the tool*
- *RPM too high: Same result, burns, not cuts.*
- *Material too dense: Wears down edge faster.*
- *Not clearing chips: Re-cutting dulls the tool faster.*
- *Improper cooling: Especially with metal, tool overheats.*

4.14 Quality Tools Are Cheaper Than Cheap Ones

Quality tools are cheaper than cheap ones, just not at checkout. A cheap tool might seem like a bargain, right up until it snaps mid-job, ruins your project (where did that metal shard go), and eats your schedule.

Invest once in solid quality tooling, and you'll spend less time fixing mistakes and more time building things.

MDF marathon? Use PCD (poly-crystalline diamond) bits.

🖵 4.15 The High Cost of Dull Tools

Saving time upfront by skipping bit checks or running dull tools usually backfires.

Dull bits do more than slow your cuts, they generate heat that can burn materials, damage your machine, and leave rough, uneven edges that demand extra sanding, re-cutting, or even scrapping the part entirely. The time you think you're saving often gets eaten up fixing mistakes or compensating for poor results.

Keeping your bits sharp isn't just about clean cuts, it's about protecting your materials, your machine's lifespan, and your production timeline.

Sharp tools cut better, last longer, and help ensure the only thing you're reworking is your next idea.

Cut a 4x8 ft sheet into a UV jig, basically adult LEGO.

📱 4.16 Success Story
Oversized Oak Door Engraving

I was commissioned to laser engrave two monumentally massive laminated solid oak doors, each far exceeding the dimensions and weight capacity of our Trotec Laser's engraving bed. This project required creative problem-solving, precision alignment, and repeatable accuracy to deliver professional-grade results on a very non-standard canvas.

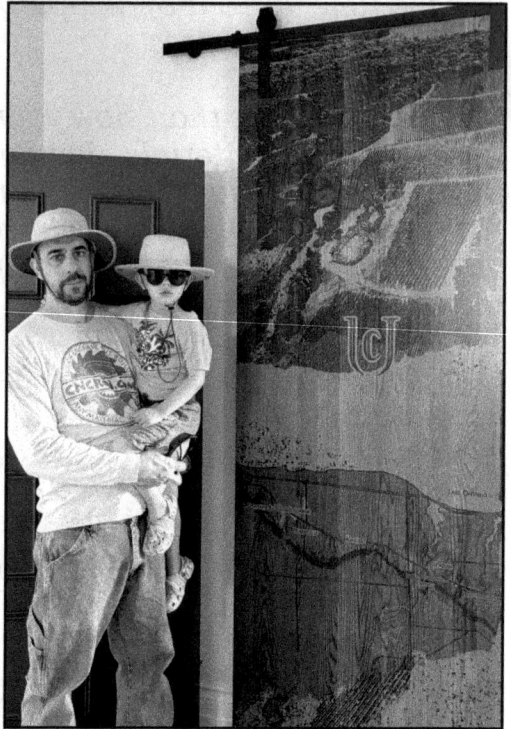

Solid laminated oak: because lifting should feel like a workout.

Project Breakdown

1. Client and Scope

The client requested deep, detailed engraving across the entire face of two oak doors, each measuring approximately 96 × 38 inches (2.4 × 1 meters) and weighing between 250 and 300 pounds (113–136 kilograms). Due to their size and mass, standard bed placement was impossible.

2. Machine & Setup Constraints

Using my Trotec Speedy 400 flexx with passthrough capability paid massive dividends here. I don't use the passthrough often, but when I do, it's incredibly profitable and saves me a lot of grief.

My CNC router lent a helping hand during this job, it held the slabs on a flat, safe surface while I exchanged them from one to the other. It can hold thousands of pounds if needed.

3. Challenges and Solutions

Weight & Support:
Heavy-duty sawhorses were positioned at both ends of the laser, supporting 2x4s that ran beneath each door. Shims and spacers ensured the engraved surface stayed level and parallel to the laser head, minimizing depth variation.

Material Inconsistencies:
Laminated oak presents density fluctuations that can affect engraving depth and contrast. I overcame this with 3–4 laser passes, each deepening the design incrementally while preserving crisp edges. Going slightly out of focus also increases charcoal buildup at the bottom of the engraving, something I intentionally used on the final pass for added contrast.

Continuous Design Across Oversized Workpiece:
Although some accuracy was lost, I had no choice but to manually advance the doors through the laser.

4. Finishing Process

Post-engraving, belt sanding was used to eliminate char and surface residue. Minor tonal shifts caused by the glue-laminated structure were smoothed out during this stage, enhancing overall visual uniformity.

Things I Had to Learn the Hard Way

Precision Setup Pays Off

Ensuring a level and stable engraving plane across the entire door minimized artifacts and depth inconsistencies. It literally took me 45-60 minutes to re-align everything after each pass.

Redundancy = Reliability

Over-engineering the support setup prevented damage and injury to both me and the machine. Redundancy in bracing adds a vital safety margin.

Move laser, burn stuff, feel productive. Repeat.

Deeper Is Safer

On hardwoods like oak, especially when sanding is expected, engraving slightly deeper preserves design clarity after post-processing.

Rethink for Repeatability

While passthrough worked here, future oversized work may benefit from modular design strategies to streamline both setup and production time, as I did with the GitHub Vault, which we'll discuss in Chapter 6's case study (p. 120).

Another One Off the Table

This project highlighted the power of CNC thinking beyond machine specs. By designing the environment around the limitations of the equipment, not the other way around, we successfully engraved massive, heavy doors with the finesse expected from smaller-scale projects.

Flexibility, foresight, and a healthy respect for gravity turned a logistical puzzle into a visual success.

Front of my Trotec Speedy 400 flexx during production.

Back of my Trotec Speedy 400 flexx during production.

🏅 4.17 Practical Takeaways from the Shop

Choose the wrong bit: You'll spend your afternoon crying over tear-out. Compression bits for plywood, upcut for clean pockets, treat your bits like specialists, not generalists.

If your material shifts mid-cut: Congrats, you've made abstract art. Screws, vacuum, tape... whatever works. Just skip the bargain-bin stuff, unless chaos is your thing.

Tiny parts love to fly: Especially into your face or your bit. Add tabs, at least 3–4 per part, unless you enjoy shop-wide scavenger hunts or shattered bits.

Even the best machine can't fix a bad design or a dirty spoilboard: Prep well, surface your bed, clean up thoroughly, check your material, verify your toolpaths, and double-check everything before you hit "Go", because a minute of prep beats an hour of fixing.

2.5D carving: because Pine deserves a little drama too.

✖ 4.18 Chapter 4 Quiz

Let's see if your CNC router skills are dialed in, or if your brain's just running on autopilot!

1. What's the primary function of a CNC router?
    A) Cutting and shaping sheet materials
    B) Cutting metals
    C) Milling 3D objects
    D) Laser engraving

2. What's a common beginner mistake in CNC routing?
    A) Using the wrong bit
    B) Jumping to the machine too early
    C) Making too many shallow passes
    D) Using too much lubrication

3. What's a key advantage of CNC over manual routing?
    A) Faster cutting
    B) Less skill needed
    C) More consistent and precise cuts
    D) Better for artistic work

4. What's true about CNC routers vs. manual routers?
    A) CNC routers can repeat cuts exactly every time
    B) CNC routers are slower
    C) Manual routers are more precise
    D) Manual routers can perform 3D carving

5. What motor type does a consumer-grade router use?
    A) Brushless motor
    B) Stepper motor
    C) Brush motor
    D) DC motor

6. What's the best tool for clean top surface cuts?
    A) Upcut bit
    B) V-Bit
    C) Compression bit
    D) Downcut bit

7. Which bit is ideal for carving smooth 3D shapes?
    A) Upcut bit
    B) Ball nose bit
    C) V-Bit
    D) Compression bit

8. What's the purpose of a vacuum table in CNC routing?
    A) To hold materials in place without clamps
    B) To ensure smooth cuts
    C) To speed up the cutting process
    D) To reduce machine noise

9. What's a sign of material shift during CNC routing?
    A) Tear-out
    B) Material chipping
    C) Inconsistent cutting depth
    D) Uneven feed rates

10. What causes "charring" in a CNC routing operation?
    A) Cutting too slow
    B) Using the wrong bit
    C) Spinning the bit too fast or feeding too slowly
    D) Incorrect bit size

11. What's the best material for fast CNC router prototyping?
    A) Aluminum
    B) Acrylic
    C) Wood
    D) Foam

12. What does "chip load" mean?
    A) Speed of the machine
    B) Material removed per tooth per revolution
    C) Diameter of the router bit
    D) Depth of the cut

13. What's the best way to avoid tear-out in CNC wood routing?
    A) Use an upcut bit
    B) Increase the RPM
    C) Use a downcut or compression bit
    D) Use a larger bit

14. What's to consider for 3D carving with a CNC router?
    A) A 3D model and ball-nose bits are required
    B) Use a flat end mill for better results
    C) Fast feed rates will provide the best finish
    D) Shallow passes should be avoided

15. What's a bad workholding method for CNC routing?
    A) Double-sided tape
    B) CNC T-track and clamps
    C) Industrial magnets
    D) Vacuum table

16. What's the term for the depth of the cut in CNC routing?
    A) Feed rate
    B) RPM
    C) Chip load
    D) Depth of cut (DOC)

17. Which of the following is a tip for routing aluminum?
    A) Use a downcut bit for clean edges
    B) Increase feed rate to avoid chattering
    C) Use lubricant and a slow feed rate
    D) Use high RPM to speed up the process

18. What controls bit speed in a CNC router?
    A) RPM
    B) Feed rate
    C) Depth of cut
    D) Chip load

19. What's the issue with cheap tape for holding material?
    A) Material shift during cutting
    B) Inconsistent cutting depth
    C) High-speed cutting
    D) Too much friction on the material

20. What's the main benefit of using tabs in CNC routing?
    A) Cleanup is easier after cutting
    B) Helps to secure material in place
    C) Improves cut speed
    D) Provides better surface finish

Thermwood CS43: carving MDF +/- molds like it owes it money.

✖ 4.19 Answer Key + Explanations

Let's find out if your answers were CNC-precise or just winging it, with explanations to back it up!

1A – CNC routers are ideal for cutting wood, plastics, and soft metals, making them versatile for fabrication.
2B – Jumping in without a clear plan often leads to wasted time, materials, money, and poor results.
3C – CNC routers use computer control for precise, repeatable cuts, unlike manual routing.
4A – CNC can make identical cuts every time; manual results depend heavily on operator skill.
5C – Consumer routers often use brush motors, which are noisy and not built for long runs.
6D – Downcut bits leave a clean top surface, great for plywood and MDF projects.
7B – Ball nose bits are perfect for carving smooth 3D contours and detailed work.
8A – Vacuum tables hold material securely without clamps, speeding up workflow.
9C – If material shifts mid-job, you'll get uneven depths and poor cut quality.
10C – Charring happens when RPM is too high or feed rate is too low, causing heat buildup.
11D – Foam is lightweight and easy to cut, making it great for fast prototyping.
12B – Chip load is how much material each cutter tooth removes per spin, it affects speed and efficiency.
13C – Downcut and compression bits reduce tear-out, keeping surface edges cleaner.
14A – 3D carving requires a digital model and ball nose bit for smooth, accurate shapes.
15C – Industrial magnets aren't reliable for holding parts firmly in CNC routing.
16D – Depth of cut (DOC) is how deep the bit goes into the material per pass.
17C – Cutting aluminum requires lubrication and slow feeds to avoid tool wear and poor finishes.
18A – RPM controls bit speed, not feed rate as feed rate directly impacts cut quality, speed, and tool heat buildup.
19A – Cheap tape often fails, causing material to shift and ruining cuts.
20B – Tabs keep parts from moving or lifting during cutting, ensuring accuracy and safety.

Don't forget to have fun with your CNCs.

Each project has lessons to be learned.

Chapter 5
CNC 3D Printing Builds Dreams
Where Ideas Take Shape One Layer at a Time

CNC 3D printing turns ideas into reality, transforming concepts into creations with just a button press (okay, after you learn how to 3D model). Build, not carve, by stacking layers into real results with impressive precision... if not exactly breakneck speed.

Custom manufacturing used to mean tracking down rare materials, wrangling tricky processes, and waiting through countless hours of hands-on work. Now? You just download a file, hit "print," and let a robot bring your vision to life in hours, or days, AI just augments the possibilities.

💬 *If CNC routers are chisels, 3D printers are tiny hot glue guns, both wielded by robots.*

My 2011 Model-T 3D model, early days! (Photo by Angus Hines.)

👎 *Thinking that your printer bed is level and flat.*

5.1 What's 3D Printing, Really?

3D printing, or additive manufacturing, if you want to sound fancy at tech parties, builds objects one whisper-thin layer at a time, guided by a digital CAD (Computer-Aided Design) file.

It feels futuristic, but the first clunky, resin-hungry machines appeared in the 1980s, moving slower than a snail on vacation.

Today, you can get a decent printer for the price of a good power tool, cranking out everything from prototypes to production parts with incredible precision while you impatiently refresh the progress app.

Different technologies offer different "flavors" of layering, depending on your taste.

Process	Strengths	Limitations & Uses
DMLS (Direct Metal Laser Sintering)	Strong, lightweight metal parts; high detail	Expensive, post-processing needed; aerospace, medical, tooling
FDM (Fused Deposition Modeling)	Cheap, easy, many materials	Weaker, rougher parts; prototypes, jigs, fixtures, education
SLA (Stereolithography)	Ultra-fine detail, smooth finish	Brittle resin, messy; dental, jewelry, miniatures, prototypes
SLS (Selective Laser Sintering)	Strong parts, no supports, complex shapes	Costly, powder handling; aerospace, prosthetics, housings, production

📥 5.2 Choosing the Right 3D Printing Tech

Choosing your 3D printing method is a bit like picking a vehicle, it depends whether you're cruising to the grocery store or launching into orbit.

If you're on a budget and just need sturdy parts fast: FDM is your trusty old pickup truck, affordable, reliable, and a little rough around the edges.

If you want jaw-dropping detail and silky-smooth surfaces: SLA is your luxury sports car, gorgeous results... if you're willing to wash, cure, and baby it afterward.

If you need strong, functional parts without fussing over supports: SLS is your heavy-duty off-road Ural 4320, capable of tackling nearly any terrain (or geometry).

If you dream of printing parts that survive outer space or major surgery: DMLS is your private jet, breathtaking, wallet-emptying, and absolutely not overkill if you have serious engineering needs.

No matter which ride you choose, the destination's the same: turning your digital dreams into physical reality, layer by glorious layer. The hardest part about 3D printing? Learning how to 3D model.

5.3 Common 3D Printing Materials

Not all materials are created equal. Some need heated beds, enclosed chambers, or industrial machines. Match your goals to what your printer can actually handle. Material "flavors" also vary:

FDM: Best for plastic prototypes and functional parts.

Resin: Ideal for detailed, artistic, or small pieces.

Metal: For extreme strength, heat resistance, and precision, not your typical garage project!

5.4 FDM Material Options

Picking FDM 3D printing materials feels like choosing pizza toppings, except some melt, warp, or refuse to cooperate.

New options seem to pop up quarterly, usually while you're still dialing in the last one. This list isn't exhaustive, but it'll help you stay afloat in the plastic sea.

Material	Strengths	Limitations & Uses
ASA (Acrylonitrile Styrene Acrylate)	UV / weather-resistant, durable	Warps like ABS; outdoor parts, auto trims
ABS (Acrylonitrile Butadiene Styrene)	Tough, heat-resistant	Strong fumes, warps easily; prototypes, enclosures
HIPS (High Impact Polystyrene)	Strong, dissolvable, good ABS support	High heat requirement, absorbs moisture; supports, lightweight parts
Nylon (Polyamide)	Strong, flexible	Moisture-sensitive; gears, hinges, mechanical parts
PC (Polycarbonate)	Extremely tough, heat / impact-resistant	High temps, warps; automotive, aerospace, tooling
PETG (Polyethylene Terephthalate Glycol)	Balanced strength, flexibility, easy to print	Can string; bottles, mechanical parts, enclosures
PLA (Polylactic Acid)	Biodegradable, easy printing	Not heat-resistant; prototypes, hobby, educational models
TPU (Thermoplastic Polyurethane)	Flexible, rubber-like	Slow, tricky to print; joints, phone cases, gaskets

5.5 Resin Material Options

Resin materials are like fancy pizza toppings, amazing when done right but not suited to every appetite.

SLA = Anchovies, precise, slow, and placed point-by-point

DLP = Pepperoni, fast and bold, with less detail but more speed

MSLA = Cheese, uniform, cost-effective, but less detailed

Material (Printer Type)	Strengths	Limitations & Uses
Bio-Resin (SLA, MSLA)	Eco-friendly, sometimes biodegradable	Weaker performance; good for sustainable prototypes, concept models
Castable Resin (SLA, DLP)	Burns out cleanly without ash	Brittle; ideal for jewelry casting, dental work, small metal parts
Ceramic-Filled Resin (SLA, DLP)	High stiffness, heat-resistant	Brittle and heavy; used for tooling, specialty engineering parts
Dental Resin (SLA, DLP)	Biocompatible, medical-grade accuracy	Expensive; used for dental models, surgical guides, orthodontics
Flexible Resin (SLA, MSLA)	Rubber-like flexibility; bendable prints	Low tear strength; best for grips, wearable parts, soft prototypes
High-Temp Resin (SLA, DLP)	High-temp resistant (>200°C / 392°F)	Brittle under stress; used for molds, fixtures, engineering parts
Standard Resin (SLA, DLP, MSLA)	High detail, smooth surface finish	Brittle; best for figurines, prototypes, jewelry models
Tough Resin (SLA, MSLA)	More durable than standard resin	Still brittle compared to FDM; used for functional prototypes, dental models

5.6 Metal Material Options

Metal 3D printing is like owning a dragon, powerful, but it demands a castle and serious ambition.

These materials offer extreme strength, heat resistance, and precision, great for aerospace, medical, and advanced manufacturing, but far from hobby-level. (DMLS = Direct Metal Laser Sintering, SLM = Selective Laser Melting)

Material (Printer Type)	Strengths	Limitations & Uses
Aluminum (AlSi10Mg) (DMLS, SLM)	Lightweight, corrosion-resistant, strong for its weight	Weaker than titanium; aerospace, automotive parts
Cobalt-Chrome (CoCr) (DMLS, SLM)	Extremely hard, wear-resistant, biocompatible	Tough to machine; dental implants, aerospace, turbines
Copper (Pure or Alloys) (DMLS, SLM)	Outstanding thermal and electrical conductivity	Harder to print; heat exchangers, electrical, rockets
Inconel (Nickel-Chromium Alloy) (DMLS, SLM)	Extreme heat and corrosion resistance	Expensive, difficult machining; jet engines, high-temp tooling
Stainless Steel (316L, 17-4PH) (DMLS, SLM)	Strong, corrosion-resistant, affordable	Heavy; medical devices, tools, molds
Titanium (Ti6Al4V) (DMLS, SLM)	Exceptional strength-to-weight, biocompatible	Very expensive, tough post-processing; aerospace, implants, racing parts
Tool Steel (H13, M300) (DMLS, SLM)	High hardness, stress and heat durable	Brittle if mishandled; molds, cutting tools, wear parts

📖 5.7 3D Printer Troubleshooting (See "It Looked Great in the Slicer..." in Appendix F)

A lot can go wrong with 3D printing. Honestly, there's enough material for an entirely separate book, How I Learned to Stop Worrying and Love My Melted Spaghetti Prints. From slicing settings that defy logic, to using the wrong filament on a printer that's doesn't want to be a 3D printer, to a bed temperature just five degrees too cold, or too hot... welcome to the rabbit hole that is 3D printing.

To save you from full-blown filament despair, I tucked a troubleshooting section into the appendices. It won't fix everything, but it'll get the ball rolling, and maybe your first layer, too.

Every printer, material, and setup has its own personality, some charming, some pure chaos. Trial and error? Still the most underrated teacher in the workshop.

My 3D castle made with my awesome 3D Systems Cube (2013)

5.8 Bridging the Digital and Real Worlds

Turning a 3D model into a physical object is where imagination collides with reality, sometimes elegantly, sometimes like a slow-motion trainwreck.

It starts with a file crafted in software like Fusion 360, SolidWorks, Blender, or 3DS Max (my go-to), and ends with something you can hold, admire, and probably drop.

Whether you're using 3D printing, CNC machining, laser cutting, or casting, the mission is simple: turn pixels into atoms.

But it's never as easy as clicking "Print." Scaling, wall thickness, tolerances, material behavior, and machine quirks all factor in, and something usually still misbehaves. That flawless digital model? Physics, heat, and gravity are waiting to introduce you to warping, chipping, or parts that technically "fit" if you convince them with a hammer.

Material choice adds its own surprises. A design perfect for lightweight PLA might become a backbreaker in solid steel. Every material brings its own strengths, quirks, challenges, and delightful "lessons."

Going from digital to physical is where real testing begins. That tiny misaligned hole or slightly warped surface? It matters now.

Prototyping is how dreams get validated, or spectacularly rewritten.

In custom manufacturing, this leap is an art built on experience, stubbornness, and trial by fire. The first version rarely wins. It's usually the third, after tweaks, fails, and a few dramatic sighs.

Bridging the gap from model to reality turns designers into problem-solvers, and gives them a lifetime of hilarious battle stories. For me, it was a massive eye opener once I got my own CNCs to make things with at *CNCROi.com*.

5.9 Strengths and Weaknesses of 3D Printing

3D printing is one of the most finicky CNC processes out there, pure wizardry one minute, an expensive way to make spaghetti the next.

It swings between moments of genius and moments that make you want to throw the printer out the window, especially when a nine-hour print fails at hour eight for no apparent reason. They have improved massively though the years though, but the gremlins still pop-out of nowhere.

Strengths

Complex Geometry: Create shapes CNC machines can't, like internal channels, honeycombs, and intricate lattices.

Material Efficiency: Use only what you need, no milling away half a block just to find your part inside.

Rapid Prototyping: Design today, print tonight, test tomorrow, ideal for fast iteration.

Low Tooling Costs: No molds, no jigs, just a digital file and some filament (or resin), keeping upfront costs minimal.

3D printing makes shapes so weird, even nature's jealous.

3D printing's way more fun when you know how to 3D model.

Limitations

Surface Finish: Layer lines are the norm, fine for prototypes, internal parts, or functional components, but not exactly showroom-ready without sanding, filling, or painting.

Mechanical Weakness: Parts are usually weaker along layer lines, anisotropy (uneven strength) is a real concern for load-bearing or snap-fit designs.

Print Speed: Detail might be "free," but time isn't, high-resolution prints can take hours (or days), especially on hobby-grade machines or for large builds.

Post-Processing Needs: Cleaning, curing, sanding, and support removal are often required before a part is truly "done," adding time and effort beyond just clicking "Print."

5.10 When to 3D Print vs. When to CNC

Sometimes you print.
Sometimes you CNC.
Sometimes you do both.
(And sometimes you just stare at a half-finished project wondering where it all went wrong, but that's a later chapter.)

The real trick is knowing when to let a 3D printer patiently build layer by layer, ideal for complex geometries, internal cavities, lightweight structures, and organic forms, and when to hand the job to a CNC machine, tearing through solid stock chip by chip with brute efficiency.

3D printing shines for intricate details, rapid prototyping, and designs that prioritize flexibility over raw strength. It brings creative concepts to life without molds, tooling, or costly rework.

But when you need durability, tight tolerances, smooth surface finishes, or faster production with rugged materials, CNC machining takes the lead.

CNC machines bring the strength and precision most 3D printers can't match, especially when it comes to metal parts and components built for real-world stress. Though, to be fair, that gap is closing *fast*. Think of 3D printing more as an augment to current capabilities instead of replacing any giving process or tool.

3D Printing is Best When

You need low-volume parts fast.
Think "I need it by Tuesday," not "I need 10,000 by Christmas."

The parts are complex, hollow, or just plain weird.
Internal channels, organic curves, lattice structures, the crazier, the better.

You're prototyping, not mass-producing.
No boss has ever yelled, "We need 500 prototypes by noon!"

CNC Machining is Best When

Parts must be super-strong and dimensionally perfect.
(CNC routers and mills don't do "close enough.")

You're making dozens, hundreds, or thousands of parts.
(CNCs love repetition, if they had a motto, it'd be: "Another one? Bring it.")

Material choices lean toward metals, hardwoods, or other tough customers.
(Spoiler: Most 3D printers hate real aluminum or wood the way cats hate bathtubs.)

The Hybrid Workflow: Best of Both Worlds

Here's a secret most seasoned manufacturers know:

You don't have to pick sides.

In fact, the smartest projects often use both.

Prototype with 3D printing: Cheap, fast, and flexible, perfect for tweaking designs while you're still in your pajamas.

Finalize with CNC machining: Once the design is locked, crank out parts with precision, strength, and repeatability.

Think of it like baking a cake: 3D printing is testing recipes, CNC is delivering the wedding cake order once you know which one doesn't taste like cardboard.

Blending both tools lets you move fast and finish strong, exploring wild ideas without wasting material, then dialing in durability when it counts. It's not about choosing sides; it's about choosing smart. In the end, the best workflow isn't 3D printing or CNC, it's knowing when to use each.

⚡ *Fast Rule of Thumb: If it looks like a skeleton, spiderweb, or coral reef, 3D print it. If it looks like a hammer, engine block, or luxury table leg, CNC it.*

⬇ 5.11 Design Tips for 3D Printing

At CNCROi.com, I usually follow a simple rule when figuring out how best to 3D print a project: make life as easy as possible for Future Me.

Most of the time, I'm also the designer of the part, which definitely helps, controlling both the pixels and the fabrication means I can quietly argue with myself about design choices before starting anything in the real world.

Having a hand in both sides lets me optimize for speed, strength, and printability.

Minimize overhangs: *Supports waste time and material.*
Orient for strength: *Layers are strong across, weak between.*
Account for shrinkage: *Especially with resins and nylons.*
Design for post-processing: *Make sanding and cleanup easy.*
Use chamfers and fillets: *Sharp corners = stress = sad prints.*

In 2013, 3D printers were slow, fussy, but they worked!

5.12 Real-World Applications

3D printing isn't just for dusty shelf figurines anymore, it's a serious (and seriously fun) tool across industries. Thanks to its additive approach, it's arguably the most versatile manufacturing method out there.

Prototyping
Need to test a design? 3D printing takes you from "wild idea" to "thing you can poke" in a day. Iterate fast, break stuff early, and fix mistakes before they get expensive.

End-Use Parts
Forget just prototypes, companies now print actual working components. Custom jigs, fixtures, replacement parts, even one-off tools that would cost a fortune to machine otherwise.

Medical
From prosthetics to surgical planning models, 3D printing helps doctors fit parts precisely to patients, rehearse procedures, and generally look like sci-fi wizards.

Automotive
Lightweight brackets, dashboards, interior bits, 3D printing saves weight and cost while letting gearheads channel their inner Tony Stark.

Aerospace
Printers now build parts that used to cost more than a sports car to machine. Satellites, rockets, and planes are flying with components fresh off a print bed.

Fun Stuff Too...
Props, cosplay armor, board game pieces, drone frames, weird gadgets you didn't know you needed, if you can dream it, you can print it... and probably forget you did by your next project.

Whether it's solving real problems or just bringing wild ideas to life, 3D printing proves that sometimes the best tool is the one that lets you invent your own.

⬚ 5.13 3D Printing: Slow Bakes and Sudden Breaks

Running a 3D printer is like running a bakery where each cupcake takes four hours... and occasionally explodes.

Material Costs: *Filament is relatively affordable; resins and powders can get pricey.* Watch for discounts!

Printer Upkeep: *Nozzles clog, beds warp, belts loosen, maintenance is part of the deal.*

Time is Money: *Long prints tie up machines (and your patience) for hours,* and days. Plan accordingly.

Intellectual Property: *Downloading a model? Make sure you're allowed to use it commercially.*

Just because you can print it doesn't mean you should, design responsibly, and print with purpose.

5.14 Future Trends in 3D Printing

Metal printing, once reserved for aerospace giants, gets cheaper every year, opening the door for smaller businesses to create strong, lightweight parts without a full-blown machine shop. FDM "metal" is slowly coming into being too.

Multi-material printing is on the rise too, allowing you to print flexible, conductive, and rigid elements all in one go, no soldering required. Large-scale printing has moved beyond keychains to houses, bridges, and boats, squeezing out structures like giant tubes of concrete toothpaste.

AI-driven design is shifting the game, generating parts that are lighter, stronger, and smarter than anything a tired human could sketch at 2 a.m.

And bio-printing? It's closing in fast, with researchers inching toward printing tissues and organs. Not quite "print your own liver" yet, but hospitals are paying attention.

☐ 5.15 Success Story
Multi-CNC Corporate Awards

At *CNCROi.com*, a corporate client asked us to create a series of custom awards blending maple hardwood, stainless steel, and 3D printed components into a single sleek, durable design. The challenge wasn't just visual, it was engineering. Each material had to fit within tight tolerances while maintaining strength, balance, and visual appeal.

This project became a real-world showcase of multi-machine synergy: a large-format CNC router shaped the wood slabs, a high-power fiber laser cut the stainless steel, a CO_2 laser engraved the wood, a low-power fiber laser marked the metal, and industrial 3D printing produced precision inserts for structural support.

Every tool played its part in turning a complex, multi-material concept into a cohesive, high-impact finished product. It all worked seamlessly too!

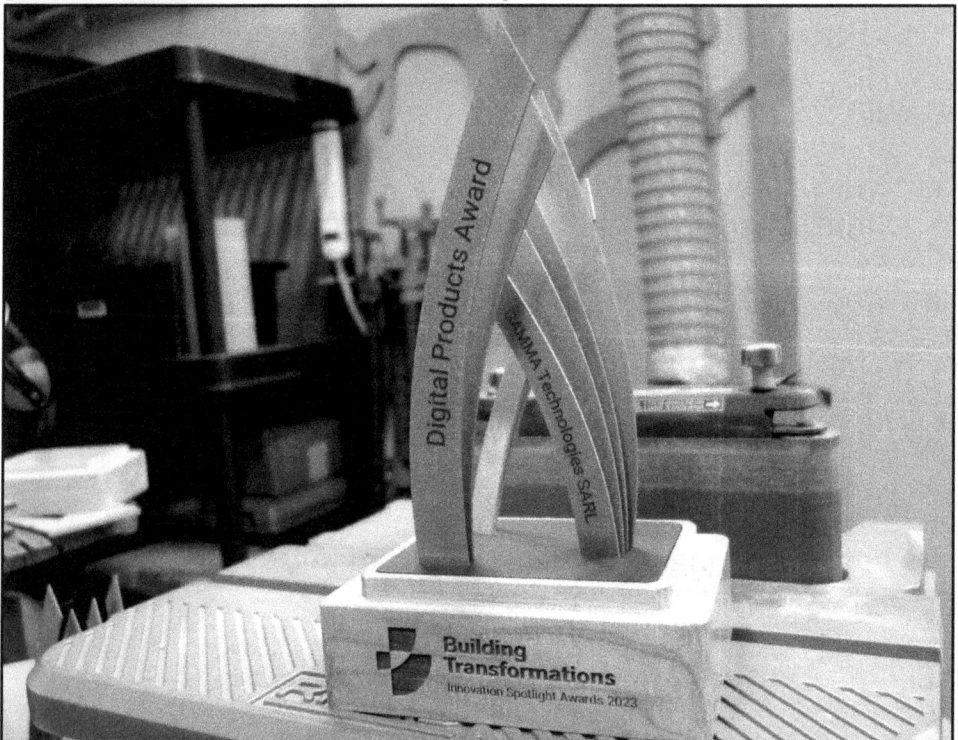

Awards can be great fun, but also quite challenging at times.

Material	Purpose	Machine Used
Maple Hardwood	Laminated and machined to create custom bases	Large-Format CNC Router, CO_2 Laser (for engraving)
16 Gauge 304 Stainless Steel	Cut into custom blade shapes and permanently etched	High-Power Fiber Laser (cutting), Low-Power Fiber Laser (etching)
3D Printed Inserts	Precision fit components to align and lock blades and bases together	FDM 3D Printer with great tolerance profile

Process Breakdown

1. CNC Routers: Building the Maple Base

Rough maple lumber was chosen for its full thickness and excellent shaping control. Boards were laminated in-house using precise glue-up and router flattening, avoiding the warping and snipe issues common with lower-grade power tool methods.

Multiple passes with varied bits reduced internal stress, keeping the bases flat and stable. Maple's light tone, hardness, crisp engraving quality, and refined appearance made it an ideal choice.

Final bases were sanded from 40 to 400 grit, creating a smooth, laser-ready finish.

2. 3D Printing: Precision Inserts

Custom 3D-printed inserts were designed to hold the stainless steel blades securely *(thanks, Walid!)*. A tight friction fit provided strength without visible fasteners or aesthetic compromise. Internal rods were printed directly into each insert, streamlining assembly and ensuring consistent precision.

Final prints were installed into the maple bases using friction and double-sided adhesive for long-term stability.

3. Laser Cutting: Stainless Steel Blades

Custom blade shapes were cut from 16-gauge 304 stainless steel using a high-power fiber laser, delivering fine detail, tight tolerances, and crisp edges with minimal heat distortion, and zero drama. The process offered the precision of a scalpel without the mess of secondary cleanup, making it ideal for clean, repeatable results.

Sharp corners were intentionally softened during the design phase to reduce handling risks while preserving the bold, dramatic aesthetic. Think "looks deadly," but safe enough to handle without cutting the winner.

Fiber laser cutting easily outperformed plasma and waterjet, offering smoother edges, less burring, and little to no post-processing. That means no sanding marathons, no mystery scorches, and no rogue metal flakes. Just blades that look sharp, feel solid, and don't threaten to bite during assembly. Small holes at the base of every blade had to be precise to match the 3D print, plasma and drilling were out.

2 kW CNC fiber laser does quick work of 16 ga 304SS.

4. Laser Engraving: Wood and Metal

The maple bases were CO_2 laser engraved on three sides, adding logos, names, and award details with crisp, high-contrast results.

The light maple tone enhanced engraving clarity without requiring paint fills.

Stainless steel blades were fiber laser etched with permanent, fade-resistant text, chosen for its durability and precision without damaging the metal surface.

Temporary jigs were fabricated to hold each blade securely during marking, ensuring consistent placement across all parts.

Post-engraving cleanup included light sanding to remove smoke residue, followed by coats of tongue oil to protect the wood and enhance its natural appearance.

Repeatable precision is limited by your material and equipment.

Challenges and Solutions

Material Matching: Maple and stainless steel were selected for their balance of aesthetics, durability, and engraving clarity.

Precision Fit: Tight tolerances were essential across 3D prints, routed pockets, and laser-cut blades, requiring careful calibration of every machine involved.

Timeline Management: In-house slab production and the coordination of multiple fabrication processes demanded tight scheduling to hit the delivery deadline.

Backup Inventory: Extra components were produced to account for potential defects or errors, a standard best practice at *CNCROi.com*.

Things I Had to Learn the Hard Way

Choose Your Materials Like You're Marrying Them

Pick and prototype materials early to avoid ugly, weak, or unengraveable surprises.

Let 3D Printing Do What CNC Can't

Modern 3D printing offers impressive precision and enables complex internal structures that complement traditional fabrication methods.

Build a Jig, Save an Hour (or Ten)

Custom jigs for temporary setups can dramatically reduce time spent on engraving and assembly.

Another One Off the Table

Over a dozen custom awards combined the warmth of hardwood, the strength of stainless steel, and the precision of 3D printing. Permanent engraving and tight tolerances ensured lasting quality and a flawless fit.

Fresh from the mill, rough, raw, and ready.

Base production? All done on my old Thermwood CS43.

🏅 5.16 Practical Takeaways from the Shop

Buy What You Need, Not What Looks Cool: FDM printers handle 90% of real-world jobs. Save the space-age resin monster for your next midlife crisis.

Design Like Gravity Exists: Overhangs and wild angles mean support structures. Ignore them, and your print will collapse faster than your weekend plans with a 5 year old.

Your First Prototype Will Be Ugly: And that's fine. Print fast, fix faster. Ugly prints teach better lessons than perfect renders ever will.

CNC + 3D Printing = Smart Workflow: Print early to tweak. CNC later to survive real-world abuse. Hybrid wins, every time.

Know the basics, press "Print," and take the credit.

Adjusted tolerances by 0.1 mm (≈ 1/254") on PLA caps using an Ender 3 Pro.

✖ 5.17 Chapter 5 Quiz

Let's see if your brain's building solid foundations, or just printing spaghetti.

1. 3D printing is what type of manufacturing process?
   A) Additive (material built up)
   B) Subtractive (material removed)
   C) Hybrid (combination process)
   D) Submersive (underwater process)

2. What decade did 3D printing first emerge?
   A) 1960s
   B) 1970s
   C) 1980s
   D) 1990s

3. Which of the following best describes FDM printing?
   A) Laser hardens resin
   B) Powder is sintered into a solid
   C) Plastic filament is extruded layer by layer
   D) Metal powder is melted into parts

4. Which 3D printing method requires curing?
   A) FDM (Fused Deposition Modeling)
   B) SLA (Stereolithography)
   C) SLS (Selective Laser Sintering)
   D) DMLS (Direct Metal Laser Sintering)

5. What's FDM's biggest weakness compared to CNC?
   A) They're too heavy.
   B) They cost too much to produce.
   C) They require hazardous chemicals.
   D) They are weak between printed layers.

6. Which 3D printing method needs no supports?
   A) SLA
   B) DMLS
   C) FDM
   D) SLS

7. Which 3D printing tech offers the most polished detail?
   A) DMLS
   B) SLA
   C) SLS
   D) FDM

8. Which material is biodegradable but heat-sensitive?
   A) PLA (Polylactic Acid)
   B) Nylon (Polyamide)
   C) PETG (Polyethylene Terephthalate Glycol)
   D) ABS (Acrylonitrile Butadiene Styrene)

9. What material is strong, yet absorbs moisture easily?
   A) PLA
   B) Nylon
   C) PC (Polycarbonate)
   D) PETG

10. Which material is tough but hard to print?
   A) ABS
   B) PETG
   C) ASA (Acrylonitrile Styrene Acrylate)
   D) PC

11. What's a major drawback of typical resin prints?
   A) They warp in sunlight.
   B) They are too flexible.
   C) They are brittle.
   D) They require laser cutting.

12. What real-world use suits resin printing best?
   A) Jewelry models
   B) Medical implants
   C) Gears
   D) Brackets for aerospace

13. Which best describes metal 3D printing?
   A) High-end, industrial-level manufacturing
   B) Cheap and easy to operate
   C) Great for mass-market consumer goods
   D) Perfect for low-stress plastic parts

14. Which metal is strong, light, and biocompatible?
    A) Aluminum
    B) Stainless Steel
    C) Titanium
    D) Inconel

15. Which of the following is **NOT** a strength of 3D printing?
    A) Speed for prototypes
    B) Complex internal geometry
    C) Low material waste
    D) High-strength metal printing with few supports

16. What's a big reason to use CNC instead of 3D printing?
    A) You need parts that are hollow.
    B) You want visible layer lines.
    C) You need precise, strong parts.
    D) You want to make a cosplay helmet.

17. Spiderweb or coral reef design, what should you use?
    A) CNC
    B) Plasma cutting
    C) Welding
    D) 3D Printing

18. Why should you use a hybrid (3DP + CNC) workflow?
    A) It makes projects slower but cheaper.
    B) It balances fast prototyping with strong final parts.
    C) It looks cooler on Instagram.
    D) It reduces file sizes.

19. In FDM printing, minimizing overhangs helps to:
    A) Save filament.
    B) Avoid supports.
    C) Speed up print times.
    D) Eliminate post-curing.

20. Watching a 3D printer work is:
    A) A great way to boost productivity.
    B) Requires a lot of stamina.
    C) Dangerous to your eyesight.
    D) Hypnotic but not productive.

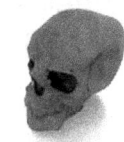

3DMarvels.com (for 3D digital files) was one of the first marketplaces I built, following WoodMarvels.com (for CNC laser / router files). Eventually, I merged them both into CNCKing.com, because why juggle two small websites when you can juggle just one slightly larger one?

*Then along came Thingiverse, basically a giant, neon-lit "**FREE STUFF**" sign. Turns out, building my CNC file empire was a lot harder when they were giving away castles for free.*

108

✖ 5.18 Answer Key + Explanations

Let's see if you built solid layers of knowledge, or if it all collapsed without enough supports!

1A – 3D printing builds parts by adding material layer-by-layer.

2C – The first commercial 3D printers appeared in the 1980s by 3D Systems.

3C – FDM melts and extrudes thermoplastic filament.

4B – SLA uses a UV laser to harden resin, creating extremely detailed parts.

5D – Layer adhesion is a common mechanical weakness in FDM.

6D – SLS builds strong parts without needing separate support structures.

7B – SLA printers are compared to "luxury sports cars", beautiful but high-maintenance.

8A – PLA is easy to print but softens easily when exposed to heat.

9B – Nylon is strong but soaks up moisture from the air.

10D – Polycarbonate is super tough but requires high printing temperatures.

11C – Standard resin prints are known for being detailed but fragile.

12A – Resin printing is ideal for detailed miniatures and jewelry.

13A – Metal printing like DMLS is advanced and expensive, not hobby-level.

14C – Titanium offers top-tier strength-to-weight and is biocompatible.

15D – Metal 3D printing often requires supports, unlike SLS, which builds in plastic powder and doesn't need them.

16C – CNC machining is unbeatable for exact, strong parts.

17D – 3D printing handles crazy geometries like spiderwebs and coral reefs.

18B – Printing for flexibility; CNC is for strength.

19B – Less overhang = fewer supports = faster, cleaner prints.

20D – Watching a printer work can be mesmerizing, and a major time sink.

It's the first time Simon has seen a 3D printer at work!

Now he knows how his Medieval Castle Walls B toy was made.

Chapter 6
CNC Laser Precision and Possibility
Details Using Energy and Motion

CNC laser systems unlock precision and possibility, using focused energy and controlled motion to create details no router or saw can match. From fine engraving to slicing thick or thin materials, laser cutters, CO_2 or fiber, deliver both accuracy and flair.

Let's explore the power behind the beam: how laser types differ, which materials they handle best, and how the right settings turn raw energy into refined results. Choosing a laser is like picking a weapon in a game, power, speed, and compatibility all count.

You'll better understand how energy becomes action, and how motion guides the beam with precision and purpose.

Memorial etched on 12 ga 304 stainless steel using a 30 W fiber.

💬 *Laser cutting is like using a lightsaber in the kitchen, amazing results, but you're in trouble fast.*

6.1 What's a CNC Laser?

A CNC laser uses focused light, so focused, in fact, that it's often invisible, like a ninja. It vaporizes, melts, or burns through material along a computer-controlled path, with zero contact. That means no wear and tear, only laser-sharp precision.

However, unlike a CNC router or 3D printer, it doesn't have a two-way feedback system on the Z-axis (height).

👎 *Neglecting to clean lenses and optics.*

6.2 CO_2 vs. Fiber Laser, Power and Precision

At *CNCROi.com*, we use both CO_2 and fiber lasers. The CO_2 excels at deep wood engraving and acrylic cutting. Meanwhile, the fiber is used for etching and cutting metals from mild steel to titanium.

Feature	CO_2 Lasers	Fiber Lasers
Wavelength	≈10.6 µm (Infrared)	≈1.06 µm (Near Infrared)
Power Range	50–4,000+ W	30–60,000+ W
Best For	Wood, acrylic, paper, leather, glass	Metals (steel, aluminum, copper...)
Cutting Speed	Slower for metals, fast for non-metals	Very fast for metals, slower for non-metals
Beam Quality	Excellent for thicker materials	Superior for precision and fine features
Efficiency	Lower (especially on metals)	Higher, especially with reflective metals
Maintenance	Requires mirrors/tube care	Minimal, solid-state design
Initial Cost	Lower at entry level	Higher, but lower running costs

While a 250+ watt CO_2 laser can technically cut metal, fiber lasers offer a finer kerf, and in the kilowatt range, they outperform CO_2 in most watt-for-watt comparisons.

6.3 Gantry vs. Galvo, Speed and Scale

Just to complicate things a little further: when it comes to marking and etching, *CNCROi.com* uses both gantry and galvo laser systems, depending on the job. Each has its own optimal production envelope.

Gantry lasers are ideal for large-format cutting and engraving where size and flexibility matter.

Galvo lasers, on the other hand, are lightning fast and made for high-precision marking in small areas, generally 4-5 inches square. Think microchips, tools, and nameplates.

Feature	Gantry Lasers	Galvo Lasers
Motion System	*Moves laser head (X/Y) over material*	*Uses mirrors to steer beam (no moving bed)*
Work Area	*Large format (up to full sheets)*	*Small format (usually <300×300 mm / 12×12")*
Speed	*Slower overall, more mechanical motion*	*Extremely fast, ideal for high-volumes*
Best For	*Cutting and engraving large parts, signs, art*	*High-speed marking, serialization, logos*
Precision	*High, but depends on motion system & optics*	*Ultra-high, excellent for fine detail*
Setup Time	*More prep (focus, fixturing)*	*Minimal, fast cycle times*
Cost (Generally)	*Lower starting cost for large format use*	*Higher per watt, but faster ROI for marking*
Depth Capability	*Good for deep engraving or thick material cuts*	*Limited depth, mostly for surface marking*

📖 6.4 Brief Laser Evolution (See "Lasers: From Lab Invention to Everyday Tool" in Appendix G)

In 1917, Einstein introduced stimulated emission, paving the way for lasers. Maiman built the first ruby laser in the 1960s. By the 1970s, lasers entered manufacturing, with CO_2 and fiber lasers disrupting industry in the 1980s. Lasers went mainstream in the 2000s, and today, fiber lasers lead in cutting, engraving, and welding.

6.5 Laser Applications Breakdown

Lasers are versatile tools, here's how each operation works and where it shines.

Operation	Description	Example Uses
Ablation	Removing surface coating / layers	Painted metals, anodized surfaces
Cutting	Full-depth material separation	Acrylic signs, wood panels, metal sheets (fiber)
Raster Engrave	Pixel-based engraving (like inkjet)	Photos, textures
Surface Marking	Color or contrast change only	Stainless steel branding
Vector Engrave	Line-based engraving	Logos, text outlines

Lasers (CO_2 and fiber) can engrave or etch logos so detailed even your grandma would be impressed, we're talking tiny text, sharp lines, and flawless finishes.

Whether you're after perfect surface textures, crisp outlines, or a brand that's permanently etched (no going back), choosing the right laser process ensures your project not only looks amazing but works like a charm, the laser equivalent of a "just right" Goldilocks moment, where power, speed, and material all align.

⬇ 6.6 Air Assist & Exhaust Systems

At CNCROi.com, I recommend skipping built-in air assist pumps and opting for an external industrial compressor. I run two 80-gallon units for full airflow control, crucial for reducing smoke and preventing flare-ups.

Exhaust is the unsung hero of any laser setup. I repurposed a dust collection blower into a dedicated fume sucker, it kicks out smoke and particulates like a no-nonsense club bouncer. No airflow? Expect scorched edges, awful visibility, irritated lungs, and probably a surprise fire show.

I keep accessories external for easier maintenance and greater flexibility.

The compressors power my lasers, Thermwood, and plasma torch, while the blower clears the shop air when I can't open the garage door across all 3 CNCs.

316 stainless vs 30 W galvo fiber, laser wins (an hour later).

6.7 Material Compatibility Guide

Not all lasers play nice with all materials, using the wrong one can turn your masterpiece into a melted mess, or worse, a toxic science experiment. Keep in mind that results vary wildly depending on your power settings, frequency, and setup. Lots of trial and error is needed for optimal results.

Material	Cut (CO$_2$ / Fiber)	Engrave (CO$_2$ / Fiber)
Acrylic	Yes / No Melts cleanly	Yes / No See lines
Aluminum	No / Yes May be reflective	No / Yes White
Glass	No / Yes Marking	Yes / No Fractures easily
Painted/Anodized Metal	No / No	Yes / Yes White
PVC	Releases toxic Chlorine gas	No
Stainless Steel	Yes / Yes	No / Yes
Wood	Yes / No Edge Char	Yes / No

6.8 Laser Parameters: Speed, Power, Frequency

Fine-tuning your laser settings is key to achieving quality results. Here's a quick look at how each parameter affects your cut or engraving.

Parameter	Description	Impact
Speed	Movement of laser head	Higher = shallower, cleaner cut.
Power	Output strength of beam	More = deeper engraving/cutting.
Frequency	Pulses per second or Hertz (PPS or Hz)	Higher = smoother edges (especially plastics).

6.9 Low-Wattage vs. High-Power Fiber Lasers

At *CNCROi.com*, our 30 W fiber galvo and gantry lasers shine at precision metal marking, but don't expect them to cut through metal, even thin sheet metal would take hours (seriously). For speed and clean kerf edges, a 1 kW+ fiber laser is the real hero. Its beam is thinner than CO_2, making it the go-to choice for fast, high-precision cutting.

Alternatively, CNC waterjets and plasma cutters are solid options for metal cutting when a high-powered fiber laser isn't available. As long as the metal is cut, who cares right?

Power Range	Typical Use	Limitation
20–100 W	*Engraving, marking, serialization*	*Not suitable for cutting*
1–2 kW+	*Cutting sheet metal fast and clean*	*Too powerful for fine engraving*

6.10 Focus & Lensing

Getting the laser focus right is like finding the perfect pair of glasses, it makes all the difference. Short lenses (1.5") are your go-to for fine-detail engraving, while longer lenses (2.5–4") allow for deeper cuts with a wider kerf.

With organic materials, you've got some wiggle room. But with metal? It feels like it's all about microns. That's why I prefer etching over annealing, etching is like the laid-back cousin who doesn't mind a little flexibility, while annealing is the picky relative obsessed with perfection.

Focus Error	Result
Too High	*Wide beam, you'll lose cutting power.*
Too Low	*Unfocused edges, poor detail, expect fuzzy edges.*
Just Right	*Sharp edges, crisp lines and peak efficiency.*

🖵 6.11 Support Your Manufacturer

Buy your lenses (and other supplies) directly from your machine's manufacturer. They might cost a bit more upfront, but it's a smart investment.

When it's time to buy another machine, rush parts, or book emergency service, that relationship can unlock serious perks, like priority support, loyalty discounts, or early access to new tech. I was the first in Canada with a Trotec Speedy 400 flexx thanks to relationships like that. The small savings from third-party parts won't help when your laser stalls mid-cut on a Friday night and you're scrambling to finish a rush job.

Let's face it, you may own the machine, but you're not escaping service calls. It's not a matter of if, but when, and usually when you're already behind.

A strong rapport with your manufacturer is the best CNC insurance policy you can buy.

Trotec Lasers: multifunctional, if you believe hard enough.

6.12 Common Laser Issues & Fixes

Laser machines can act up in all sorts of weird ways (like any CNC, really), but most problems vanish faster than smoke in an exhaust fan, just run a few quick checks and you're back in business.

Problem	Cause	Solution
Poor Cuts	Dirty lens, wrong focus	Clean optics, check height
Incomplete Engraving	Low power or misfocus	Re-focus and adjust settings
Ghosting	Loose mirrors, bed not level	Check alignment and re-level
Burnt Edges	No air assist	Add / clean air line

6.13 Design Prep Checklist

Use hairline strokes for vector cuts and set raster engravings to 300–600 DPI. Convert all text to outlines to avoid font issues, and color-code your design elements for easy operation ID (e.g., red for cuts, blue for engraving).

6.14 Summary of Laser Types

CO_2 **Lasers:** Perfect for non-metals like wood, acrylic, and leather. Handles large or detailed jobs depending on setup (flatbed vs. galvo).

Fiber Lasers (Standard): Best for cutting or marking metals up to ≈6 mm (1/4").

kW-Range Fiber Lasers: High-powered (1+ kW) and built for fast, efficient metal cutting.

MOPA Fiber Lasers: Great for fine control and color marking on metals like titanium and stainless steel.

UV Lasers: Used for cold marking and ultra-fine detail on delicate materials (glass, plastics, electronics).

📱 6.15 Success Story
The GitHub Arctic Code Vault

In 2020, GitHub launched a visionary initiative to preserve humanity's collective open-source knowledge for future generations: the *GitHub Archive Program*.

At the heart of this digital time capsule was the Arctic Code Vault, a long-term preservation project stored deep within the permafrost of Svalbard, Norway, just 1,000 kilometers (620 miles) from the North Pole. Built to survive the test of time (and maybe even the next Ice Age), the vault was designed to encode and protect vital open-source code for centuries to come.

But turning digital dreams into something physically durable isn't a job for chance, or a USB stick. That's why GitHub turned to *CNCROi.com* to transform their pixels into a permanent, tangible archive.

Lemarchand's box energy, without the workplace violations.

Project Breakdown

1. Material Selection

CNCROi.com was tasked with building an ultra-durable vault. After consultation, 304 stainless steel was selected for its corrosion resistance and cost-effectiveness.

The structure used 3-gauge (≈ 6.4 mm / 1/4") steel and 12-gauge (≈ 2.5 mm / 1/8") for the outer etched plates.

Even the hinges were custom-machined to ensure long-term reliability.

Proof our shelving holds up: we didn't just aim for 'safe', we went straight into overkill. Check out the extra weight plates beside and behind me.

2. Fabrication Process

The vault was built for precision and permanence, using CNC fiber laser cutting and stainless GMAW welding. Afterward, fiber laser etching added permanent markings, designed to outlast lamacoid tags long after they fade.

3. Pandemic Challenges

COVID-19 caused major supply disruptions in 2020, but *CNCROi.com* overcame them through creative sourcing, remote coordination, and an unwavering focus on quality.

4. Final Deployment

The finished vault was installed 250 meters underground in a decommissioned coal mine in Svalbard, a geologically stable site designed to preserve its contents for millennia.

Things I Had to Learn the Hard Way

Material Matters

Stainless steel wasn't just for looks, it was chosen for its durability and tamper resistance in extreme environments.

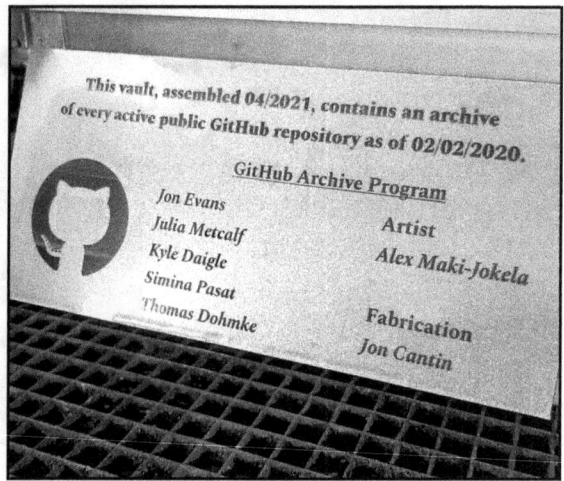

This vault, assembled 04/2021, contains an archive of every active public GitHub repository as of 02/02/2020.

GitHub Archive Program

Jon Evans
Julia Metcalf
Kyle Daigle
Simina Pasat
Thomas Dohmke

Artist
Alex Maki-Jokela

Fabrication
Jon Cantin

Design for Permanence

Fiber laser etching offered longevity far beyond traditional labels or engravings, and we do plenty of those at my shop. We initially considered using 316 stainless steel for even greater durability, but due to cost, we went with 304.

Expect the Unexpected

Global events like a pandemic can derail even the best-laid timelines, *CNCROi.com* was as flat-footed and surprised as any other business. Resilience and agility are key to staying on track. For this project, the welding shop we used had to shut down for two weeks after all their welders got infected. COVID-19 spread in a weld shop? One heck of a bug!

Custom Everything

Off-the-shelf wasn't an option, even the hinges had to be custom-built to meet the vault's exact performance and dimensional requirements. Those doors were *HEAVY*!

Another One Off the Table

This wasn't just metal, it was legacy. *CNCROi.com* helped GitHub preserve digital knowledge by pairing modern manufacturing with long-term vision. Built to survive time and crisis, it proves that custom fabrication can meet future challenges. I even asked if we could make a spare... but the quality was too good to justify it, still, a guy can dream!

Weld prep made easy: shiny steel, sticky notes, zero guesswork.

🎖 6.16 Practical Takeaways from the Shop

Nail the Focus: A tiny focus misstep can turn clean cuts into blurry messes or scrap the whole piece. It's like threading a needle while riding a bike, precision is everything.

Test New Materials: Treat unfamiliar materials like the new kid at school, always run a test first. What works on one might totally fail on another, and a quick trial now saves big headaches later.

Engraving Takes Time: Engraving is slower than cutting. Think of it as painting a portrait versus rolling paint on a wall, detailed, deliberate, and absolutely worth the time.

Respect the Laser: Lasers are unforgiving. Always have proper safety gear and stay alert. I've known shop owners (including one who shares my son's name) burn down their shops by running jobs unattended overnight.

❌ 6.17 Chapter 6 Quiz

Fire up your brain like a laser, time to test your knowledge!
No pressure... just your ego on the line.

1. What's the primary benefit of using a CNC laser?
   A) Reduced material costs
   B) Increased speed and precision
   C) Reduced waste
   D) Easier design software integration

2. Which type of laser is typically used for cutting metal?
   A) CO_2 Laser
   B) UV Laser
   C) Fiber Laser
   D) MOPA Fiber Laser

3. What wavelength does a CO_2 laser typically operate at?
   A) ≈10.6 µm
   B) ≈1.06 µm
   C) ≈0.5 µm
   D) ≈0.9 µm

4. Which material is best for fiber laser etching?
   A) Wood   B) Acrylic
   C) Stainless Steel   D) Glass

5. Which material releases toxic gas when laser-cut?
   A) PVC   B) Wood
   C) Acrylic   D) Aluminum

6. What does 'frequency' in laser parameters affect?
   A) Depth of cut
   B) Beam focus
   C) Edge smoothness
   D) Speed of engraving

7. Best power range for fiber laser etching or marking?
   A) 20–100 W
   B) 500–750 W
   C) 1–2 kW
   D) 6+ kW

8. What should you do if your laser is cutting poorly?
    A) Replace the laser.
    B) Clean the lens and check the focus.
    C) Change the material.
    D) Increase the power output.

9. What's the best lens choice for deep laser cutting?
    A) 1.5" lens   B) 1" lens
    C) 5" lens   D) 2.5–4" lens

10. Which material requires a CO_2 laser for engraving?
    A) Stainless Steel   B) Wood
    C) Aluminum   D) Copper

11. What's a typical application of a 1–2 kW fiber laser?
    A) Engraving fine details
    B) Cutting sheet metal
    C) Marking plastics
    D) Etching anodized aluminum

12. Which laser is best for plastics, glass, and electronics?
    A) CO_2 Laser   B) Fiber Laser
    C) MOPA Fiber Laser   D) UV Laser

13. Which laser is best for high-speed metal cutting?
    A) CO_2 Laser   B) 30 W Fiber Galvo Laser
    C) 1–2 kW+ Fiber Laser   D) UV Laser

14. What happens if you laser cut without air assist?
    A) Ghosting
    B) Burnt edges
    C) Poor material alignment
    D) Incomplete engraving

15. Which material is unsafe for laser cutting?
    A) Acrylic   B) Stainless Steel
    C) PVC   D) Wood

16. What should you check if engraving is incomplete?
    A) Laser power or focus   B) Air assist settings
    C) Material type   D) Workpiece temperature

17. Which laser is used for color marking on metals?
    A) Fiber Laser   B) MOPA Fiber Laser
    C) CO_2 Laser   D) UV Laser

18. What does the term 'ablation' refer to in laser cutting?
    A) Full-depth material separation
    B) Pixel-based engraving
    C) Line-based engraving
    D) Removing surface coatings

19. What's an ideal material for cutting with a CO_2 laser?
    A) Stainless Steel   B) Aluminum
    C) Wood   D) Copper

20. Which laser setting ensures sharp, detailed engraving?
    A) Proper focus
    B) High speed
    C) Low power
    D) High frequency

Lamacoids: the MVPs of industrial tagging.

✖ 6.18 Answer Key + Explanations

Time to focus in, here's the laser-sharp answer key with all the clarity you were looking for!

1B – *Lasers combine speed and precision, making them ideal for detailed custom fabrication.*
2C – *Fiber lasers are best for cutting metals; CO_2 excels at non-metals like wood and acrylic.*
3A – *CO_2 lasers operate at ≈10.6 μm, optimized for non-metal cutting and engraving.*
4C – *Fiber lasers etch stainless steel cleanly thanks to their high power and pinpoint precision.*
5A – *Cutting PVC releases toxic chlorine gas, just don't.*
6C – *Higher frequency (more pulses per second) smooths edges, especially in plastics.*
7A – *20–100 W fiber lasers are ideal for marking and engraving metals.*
8B – *Poor cuts usually mean dirty lenses or bad focus, both easy fixes.*
9D – *Longer lenses (2.5–4") allow deeper cuts but widen the kerf.*
10B – *CO_2 lasers are your go-to for engraving wood; fiber won't cut it here.*
11B – *A 1–2 kW fiber laser is built to cut sheet metal quickly and cleanly.*
12D – *UV lasers are best for delicate materials, they "cold mark" without heat damage.*
13C – *High-power fiber lasers (1–2 kW+) are made for fast, industrial-grade metal cutting.*
14B – *No air assist = burnt edges, poor results, and smoky regrets.*
15C – *Never cut PVC, it releases hazardous chlorine gas.*
16A – *Incomplete engraving usually means low power or a misaligned focus.*
17B – *MOPA fiber lasers offer fine control and can color-mark metals like titanium.*
18D – *Ablation is surface-layer removal (like paint or anodizing) without harming the base.*
19C – *CO_2 lasers are ideal for organic materials like wood but not metals.*
20A – *Sharp, detailed engraving starts with perfect focus, no exceptions.*

30 W Galvo vs 30 W Gantry fiber laser, same power, same results.

CNC plasma-cut and fiber-etched 12-gauge 304SS slats.

Chapter 7
CNC Plasma Ionized Firepower
Vaporizing Metal Into Shop Smog

CNC plasma brings ionized firepower to the workshop, slicing metal into sparks, smoke, and smog with brutal precision. It's not just fast, it's raw energy cutting steel like a railgun through armor plate.

From handheld torches to CNC tables, this covers how air and voltage create a jet of controlled disintegration. It looks at systems from basic to HD setups that rival laser accuracy, balancing speed, power, and cost.

We'll track the shift from manual to automated cuts, explain why HD plasma is worth it for some applications, and compare it to lasers to see who rules the floor.

If you want fast, fiery results, and don't mind smoke, plasma's your CNC of choice when it comes to metal.

💬 *Plasma cutting is like directing a focused electric storm to carve intricate shapes from metal.*

7.1 What's Plasma Cutting?

Plasma cutting is like turning your torch into a mini lightning bolt. It uses a superheated jet of ionized gas (plasma) to melt and blast away material from a workpiece. The process starts when an electrical arc forms between the electrode and the grounded material, ionizing the gas and transforming it into plasma.

This plasma stream carries the arc to the surface, generating enough heat to melt metal almost instantly. As the gas blows the molten metal away, you're left with a clean, precise cut and minimal slag, like a highly-focused cleanup crew for your steel. It's my favorite CNC process.

👎 *Ground it, like you would do if you were welding it or watch your arc go on a wild adventure.*

CNC Plasma Cutting is HOT, LOUD and FUN!

7.2 Plasma Cutting: Manual vs. CNC vs. High-Def

Manual plasma is the shop's duct tape, quick, messy, and always handy when precision takes a back seat to speed. Great when you just need metal separated *NOW!*

CNC plasma adds brains to the brawn, turning chaos into clean, repeatable cuts. It's your go-to when parts actually need to fit, align, and show up more than once.

High-definition plasma is the VIP, tighter arcs, razor-sharp edges, and enough precision to make lasers sweat. It's fancy, it's fast, and it gets the job done beautifully.

Each has its place, depends whether you're patching a fence or making art, and whether "good enough" is actually good enough.

Feature	Manual Plasma	CNC & HD Plasma
Control	Hand-guided	CNC-controlled (HD adds more precision & automation)
Cut Quality	Operator-dependent	Consistent (HD offers near-laser quality)
Thickness Range	Up to 38 mm (1.5")	6 mm to 50+ mm (1/4" to 2+")
Speed	Variable, depends on operator	Fast and repeatable (HD optimizes speed / quality)
Cost	Variable, depends on operator	Higher cost (HD = premium setup)
Gas Type	Typically compressed air	Air, O_2, N_2, or mixed gases (HD = precise blends)
Best Use	Quick jobs, fieldwork, small shops	Repeat parts, prototyping, or industrial cutting

⊡ 7.3 When Steel Gets Tough, Plasma Gets Going

At CNCROi.com, plasma cutting is our go-to for fast prototyping and quick-turn jobs that don't require ultra-fine detail. That's what laser and waterjet are for.

My manual plasma torch handles the rough work, trimming oversized steel parts on the fly, before more precise processes take over. It's fast, flexible, and perfect for on-the-spot adjustments.

Laser and waterjet are better suited for polished, high-volume production runs but their advantages aren't always required. When it comes to cutting complex shapes, especially in thicker mild steel where lasers slow down or costs climb, CNC plasma steps in to do the heavy lifting with speed, accuracy, and just enough flair to keep things interesting. Each CNC has their optimal use case.

7.4 Plasma Cutting Applications

Plasma cutting is incredibly versatile, from quick trims to intricate geometries, it's a go-to tool across industries, from heavy fabrication and auto repair to custom signage and even sci-fi effects in film. Its ability to cut thick metal quickly with minimal setup makes it a shop essential.

Operation	Description	Example Uses
Rough Cutting	Handheld torch for basic shapes	Large brackets, structural components
Precision Cutting	CNC-controlled for intricate designs	Metal signage, complex brackets
Bevel Cutting	Angled cuts using specialized torch heads	Weld preparation, creating gussets
Piercing	Initiating cuts within the material	Creating bolt holes, embedded features

Custom firepit, fresh from the plasma spa.

7.5 Material Compatibility Guide

Plasma cutting shines on conductive metals, if it can carry a current, it's fair game. But performance depends on the metal's properties and thickness.

It's like a circus act: mild steel is the strongman, tough, steady, and unfazed by heat, while thin aluminum is the tightrope walker, one misstep, and things get messy.

Material	Plasma Cutting Suitability	Notes
Mild Steel	Excellent	Clean cuts with minimal dross.
Stainless Steel	Good	May require post-cut cleanup to remove dross.
Aluminum	Moderate	Prone to dross buildup; requires fine-tuned settings.
Galvanized Steel	Good	Effective cutting; **ensure proper ventilation.** **Zinc Oxide = metal fume fever**
Copper / Brass	Limited	High reflectivity and thermal conductivity pose challenges.
Non-conductive Materials	Not suitable	Plasma cutting requires electrically conductive materials.

7.6 Key Plasma Cutting Parameters

Adjusting plasma cutting parameters is like running a circus, everyone needs to be in the right place at the right time, or the whole show goes sideways.

Too much amperage? The fire breather scorches the tent. Too slow on travel speed? The juggler drops everything.

But get the balance right, and it's a standing ovation: smooth edges, clean cuts, and no drama with the clowns.

Parameter	Description	Cutting Impact
Amperage	Controls the electrical arc intensity	Higher amperage enables cutting thicker materials.
Torch Height	Distance between torch and workpiece	Critical for achieving desired cut quality.
Cutting Speed	Rate at which the torch moves along the cut path	Affects edge quality and dross formation.
Gas Type	Type of gas used in the plasma stream	Influences cut cleanliness and precision.
Copper / Brass	Limited	High reflectivity and thermal conductivity pose challenges.
Non-conductive Materials	Not suitable	Plasma cutting requires electrically conductive materials.

While every plasma setting plays a part, torch height control is the one most likely to hijack your project if ignored. Think of it like spray painting, hold the can too far, and the coverage is weak and patchy; too close, and you get drips, splatter, and a mess.

Getting the torch height just right is essential for clean, consistent cuts. Too high, and the arc weakens, penetration suffers, and your edges develop an unwanted bevel. It's like trying to toast bread from across the room, dramatic, but useless.

Too low, and you're in a whole different circus act: the nozzle grinds into the metal, dross builds up, and the torch might start slam-dancing with your workpiece. Not ideal for tool longevity, or your sanity.

But when you find that ideal distance, the magic happens: smooth, crisp cuts, minimal cleanup, and accurate dimensions that save time and rework.

7.7 Manual vs. CNC vs. Oxy-Acetylene Cutting

I was first introduced to oxy-acetylene cutting during my time in the two-year welding program at Niagara College, at the bright young age of 47, thanks to Dave.

At first, it was intimidating. Fire, pressure, and sparks don't exactly scream "comfort zone." But once I got the hang of it and understood how it all worked, it became clear why heavy industry leans on it so much: it's simple, powerful, and highly effective.

Oxy-Acetylene Cutting

Oxy-acetylene cutting works by mixing oxygen and acetylene gas to create a high-temperature flame that melts metal. An additional oxygen stream then blows the molten material away, accelerating the natural oxidation process. It's a straightforward but powerful method, more cost-effective and portable than plasma in many cases.

Advantages:
- Portable and versatile, ideal for field work or when electricity isn't available.
- Handles thick metal well (up to ≈100 mm or 4 inches).
- Lower upfront cost compared to CNC-based systems.

Disadvantages:
- Slower cutting speed than plasma.
- Less precise, with rougher edge quality.
- Higher gas consumption means higher operating costs.
- Requires more operator skill for consistent results.

Oxy-Acetylene vs. Plasma Cutting:

Plasma cutting offers fast, clean results on thinner materials, typically up to 50 mm (2 inches) thick, making it a solid choice for sheet metal and structural fabrication. It also creates sharper edges with less slag, which cuts down on the need for grinding or post-processing and saves time during finishing.

Hot rolled steel, just waiting on its enamel beauty treatment.

When a client approached *CNCROi.com* for a rugged yet refined outdoor sign, I delivered a custom cottage piece made from 12-gauge (≈2.7 mm or ≈1/8") mild steel.

Finished with black spray paint for a reflective sheen, the project balanced both form and function, CNC plasma cutting brought durability and design flexibility together.

Process Breakdown

1. Material Selection and Importance

Mild steel was chosen for its balance of strength, cost, and ease of fabrication.

While stainless steel offers better corrosion resistance, mild steel met the project's needs and budget more practically for an outdoor setting.

CNC plasma: where the kerf tells the whole story.

2. Cutting with Precision

CNC plasma cutting enabled high-detail execution with minimal heat distortion. Internal features were cut before the outer profile to maintain tight tolerances throughout the process. *CNCROi.com* uses an air-cooled setup instead of a water table to avoid flash rust and stagnant water issues. Water tables have their place, just not in my shop.

3. Built for the Elements

To ensure long-term durability, enamel paint was sprayed on all surfaces. Thicker sections were added in stress-prone areas, like around the fishing rod graphic, to prevent warping or breakage over time.

4. Mounting Made Simple

Mounting holes were integrated directly into the design, eliminating the need for extra drilling and preserving the clean finish.

Things I Had to Learn the Hard Way

Design with Strength in Mind

Don't just think about how it looks, think about how it will survive the elements and stand the test of time.

Cut Smart, Not Fast

Cutting order and system choice directly affect the final product's precision, strength, and longevity.

Don't Overlook the Invisible

Small changes, like adding even a fraction of thickness or including extra support bridges, can significantly boost structural integrity, without changing the design's appearance.

Another One Off the Table

By combining thoughtful material selection, strategic reinforcement, and precision CNC plasma cutting, the result wasn't just a sign, it was a durable, custom-built expression of craftsmanship.

Hand torch dreams, CNC plasma reality.

🎖 7.9 Practical Takeaways from the Shop

Avoid these common mistakes: Poor grounding, incorrect torch height, wrong speed, or using the wrong gas can ruin your cuts. And remember, plasma only works on conductive materials, so don't waste your time on plastic or wood unless you cover its surfaces with metal.

HD plasma is the laser's cooler cousin: It delivers precision and speed, especially on thicker or more complex cuts. Think of it as the laser's sharp, stylish sibling, but one who still knows how to have fun.

Torch height control is everything: Too high and your cut's sloppy; too low and you risk torch damage. Like a tightrope act, it's all about balance.

Air cooling beats water tables: It prevents flash rust and avoids stagnant water issues, but without proper airflow, it's like trying to cool off with a hairdryer in a sauna.

Plasma cutting delivers, but only if you respect its quirks. Nail the setup, choose the right materials, and control your torch height.

❌ 7.10 Chapter 7 Quiz

CNC plasma trivia time, no dross allowed!

1. How does plasma cutting remove material?
    A) Friction and pressure
    B) A focused laser beam
    C) Chemical reaction between gases
    D) Ionized gas melts and expels material

2. Which material isn't suitable for plasma cutting?
    A) Mild steel
    B) Copper
    C) Plastic
    D) Aluminum

3. What's the role of the electrical arc in plasma cutting?
   A) It guides the torch head.
   B) It forms the plasma by ionizing the gas.
   C) It maintains gas pressure.
   D) It blows away dross.

4. What's a common beginner mistake in plasma cutting?
   A) Using too much nitrogen
   B) Improper grounding
   C) Cutting too slowly
   D) Cutting stainless steel with air

5. How is HD plasma different from regular CNC plasma?
   A) Ability to cut wood
   B) Use of handheld torches
   C) Superior cut quality near laser-level precision
   D) Smaller footprint

6. Which cuts mild steel under 2" fastest and cleanest?
   A) Manual plasma cutting
   B) CNC plasma cutting
   C) Oxy-acetylene cutting
   D) Hacksawing

7. Which parameter affects material thickness in cutting?
   A) Cutting speed
   B) Torch height
   C) Amperage
   D) Gas flow rate

8. Why is torch height control critical?
   A) It reduces gas consumption.
   B) It keeps the arc stable.
   C) It maximizes voltage.
   D) It ensures quality cuts with minimal dross.

9. Which gas is commonly used in basic plasma cutting?
   A) Helium
   B) Oxygen
   C) Acetylene
   D) Compressed air

10. What's *the* drawback of oxy-acetylene over plasma?
    A) Lower precision and slower speed
    B) Higher electricity cost
    C) Cannot be used outdoors
    D) Cannot cut anything over 1 inch

11. Which material is hazardous when plasma cutting?
    A) Stainless steel
    B) Galvanized steel
    C) Brass
    D) Copper

12. Why is a water table not used with my plasma cutter?
    A) It's illegal to use one for cutting metal.
    B) It's more expensive than air cooling.
    C) It avoids stagnant water and rusting issues.
    D) It decreases cut quality.

13. What was the steel thickness in the cottage sign project?
    A) 2.7 mm (12-gauge) ≈ 7/64 inch
    B) 3 mm (11-gauge) ≈ 1/8 inch
    C) 1.5 mm ≈ 1/16 inch
    D) 6 mm (3-gauge) ≈ 1/4 inch

14. What happens if the torch is too low for plasma cutting?
    A) Cuts are clean and fast
    B) Nozzle damage and excessive dross
    C) Poor arc formation
    D) Torch loses power

15. Which application benefits from plasma bevel cutting?
    A) Cutting holes in plastic
    B) Making fine jewelry
    C) Weld preparation for thick steel
    D) Decorative wood signs

16. Why can't plasma cut non-conductive materials?
    A) It doesn't cut hot enough.
    B) It depends on electrical conductivity.
    C) Gas pressure is too high.
    D) It uses oxygen exclusively.

17. Why add mounting holes to a plasma-cut design?
    A) Increases weight for strength.
    B) Simplifies installation and maintains finish.
    C) Speeds up painting.
    D) Makes future repairs easier.

18. What's a key advantage of manual over CNC plasma?
    A) Portability and flexibility
    B) Cheaper material use
    C) Cleaner edges
    D) More precision

19. When is CNC plasma cutting ideal for business use?
    A) When maximum thickness is required
    B) For hobby-level metal artwork
    C) For cutting plastic signs quickly
    D) For medium-to-high production

20. What key lesson did the case study highlight?
    A) Always paint before cutting.
    B) Skip pre-drilled holes to save time.
    C) Design for strength and think long-term.
    D) Avoid using mild steel outdoors.

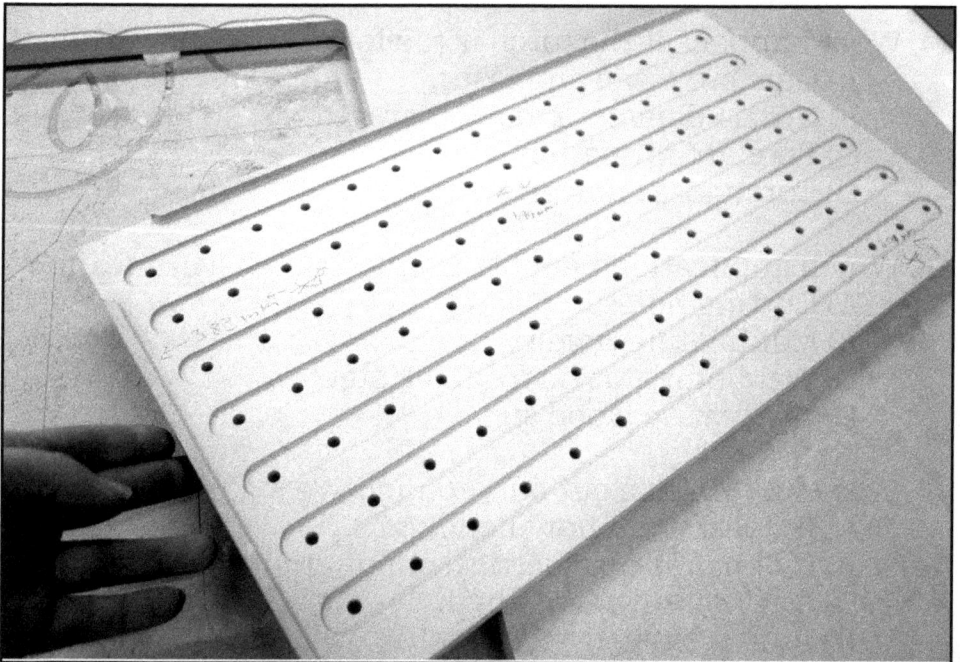

MDF makes perfect UV printer jigs, cheap, flat, and faithful.

✖ 7.11 Answer Key + Explanations

Answer key time, let's see if you cut through the confusion like a plasma torch and know why (or why not)!

1D – Plasma cutting uses ionized gas to melt and remove metal, not force or chemicals.
2C – It only works on conductive materials, plastic isn't one.
3B – The arc ionizes the gas, turning it into plasma that conducts electricity and cuts metal.
4B – Poor grounding causes unstable arcs, bad cuts, and potential torch damage.
5C – HD plasma offers tighter control and near-laser precision.
6B – CNC plasma delivers fast, clean cuts on thinner metals, outperforming manual and oxy-fuel.
7C – Higher amperage increases arc energy, letting you cut through thicker materials.
8D – Correct torch height minimizes dross and ensures clean, consistent cuts.
9D – Compressed air is the go-to gas, especially in basic or portable systems.
10A – Oxy-fuel is slower and less precise than plasma, especially on thin materials.
11B – Galvanized steel gives off toxic zinc fumes when cut, always ventilate.
12C – CNCROi.com avoids water tables to prevent rust and stagnant water issues.
13A – The cottage sign was made from 12-gauge mild steel (≈ 2.7 mm or 7/64"). Cheap and easily sourced gauge.
14B – If the torch is too low, you'll get nozzle damage, excess dross, and poor cut quality.
15C – Bevel cuts are commonly used to prep thick steel for welding.
16B – Plasma cutting relies on electrical conductivity, no current, no cut.
17B – Built-in mounting holes simplify installation and protect the final finish.
18A – Manual plasma is ideal when portability and flexibility matter more than automation.
19D – CNC plasma shines in medium-to-high volume production with repeatable, accurate cuts.
20C – CNCROi.com's project emphasized strength, longevity, and real-world usability.

Steel meets plasma... steel loses, beautifully.

Blanks by plasma, finesse by laser, teamwork!

Chapter 8
Welding for Fabrication Success
Where Sparks Meet Skill, and Creativity Ignites

Welding isn't just about fusing metal; it's about turning a puddle of molten steel into something that brings your design dreams to life. Whether you're working with massive metal plates or fine-detail projects, welding can be your secret weapon, as long as you remain grounded.

We'll explore different welding methods, the craft of making clean, strong joints, and why every good welder knows that the right technique matters more than simply cranking up the heat. Strap on your helmet, zip up that PPE, we're about to jump into a skill that's more thrilling than a hole-in-one!

💬 *Welding is the pursuit of perfection, shaping raw amps into precise, lasting bonds, one spark at a time.*

Who needs the Sun when you've got a welding arc?

8.1 What's Welding?

Welding is an essential skill in custom metal fabrication, where you create strong, permanent joints by melting and bonding materials together. Think of it as making a metal friendship bracelet, except instead of string, you're using molten steel. With techniques like GMAW, GTAW, and SMAW, MCAW, you control the strength, appearance, and efficiency of your final product.

Learning to weld is a blast. Every arc strike plunges you into an exciting mix of heat, metal, and precision. As your skills grow, you'll appreciate the hands-on nature and the instant feedback that comes with every bead.

It might feel tricky at first, but once you find the rhythm, welding becomes both fun and rewarding. Bonus: it's great for hand-eye coordination. Welding is like golf, only with more sparks and radiation while chasing a hole.

A solid stance, something all my fantastic welding instructors emphasized, made all the difference for me at age 47. It helped eliminate shaky movements and sharpen my technique. This chapter is just a taste, welding is a deep and exciting field worth exploring.

👎 *Overdo the gas and you're welding with a geyser.*

8.2 Welding is Everywhere

From the frame of your car to the chair you're sitting on, from the streetlight outside to the bridges you drive across, welding quietly holds the world together.

It's behind the scenes in appliances, buildings, bikes, playgrounds, pipelines, even art installations. Those clean seams and sturdy joints? That's metal fused with heat and precision by skilled hands.

Welding is easy to overlook, but without it, most of the structures and tools we rely on wouldn't exist, or at least wouldn't last long.

8.3 Core Welding Processes

Choosing the right welding process is like picking between a hammer and a chainsaw, the wrong choice can lead to disaster. Nail it, and you're the welding wizard you were meant to be. Niagara College was my Hogwarts, two years of sparks and craft, chasing that perfect bead.

Process	External Shielding Gas?	Typical Use
FCAW-G Flux-Cored Arc Welding (Gas)	Yes	Cleaner welds, controlled environments
FCAW-S Flux-Cored Arc Welding (SelfShielded)	No	Outdoor work, quick tacking
GMAW Gas Metal Arc Welding (MIG / MAG)	Yes	Fast production welds, clean finish
GTAW Gas Tungsten Arc Welding (TIG)	Yes	Precision, cosmetic, or aluminum welds
MCAW Metal-Core Arc Welding	Yes	High deposition rate, thick materials
SMAW Shielded Metal Arc Welding(Stick)	No	Fieldwork, repairs, general fabrication

⚠️ *Welding processes are constantly evolving, so it's important to stay up to date, this is not a stagnant industry!*

8.4 MCAW vs GMAW

Metal-Cored Arc Welding (MCAW) vs. Gas Metal Arc Welding (GMAW), it's like choosing between a pickup truck and a race car.

One's built for speed and impact, the other's your reliable all-rounder.

Higher Deposition Rates

MCAW lays down weld metal like it's on a caffeine buzz, ideal for thick steel and big jobs where time is money.

Better Fusion on Dirty Steel

Got grime? MCAW doesn't mind. It handles less-than-pristine metal better than GMAW, meaning less scrubbing and more welding.

Less Spatter, More Swagger

MCAW makes cleaner welds with less mess, so you spend less time grinding and more time pretending you meant to weld that well.

Faster Travel Speeds

It moves faster across the joint, like a welder with lunch break on the brain, without sacrificing quality.

Tougher Welds

MCAW welds come out stronger and more impact-resistant, perfect for structural work or anything that takes a beating.

Downside?

MCAW requires more advanced equipment, precise parameter control, and clean shielding gas, making it best suited for controlled shop environments. Under these conditions, it delivers high deposition rates, deep penetration, and consistent weld quality, which makes it ideal for structural and high production fabrication. MCAW stands out as the high performance option.

GMAW remains the most versatile choice for everyday welding. Its simpler setup, broad material compatibility, and forgiving operation allow it to perform reliably across many job sites and applications. For routine work, GMAW continues to be the standard.

⬇ 8.5 Welding Without Gas Tanks

At CNCROi.com, we skip the gas to save space, cut costs, and boost safety. By using Self Shielded Flux-Cored Arc Welding (FCAW-S) and Shielded Metal Arc Welding (SMAW), we get strong, reliable welds without the gas tanks.

These methods are perfect for compact shops or fieldwork where flexibility is key. While FCAW-S can be messier than GMAW, its speed and power, especially on carbon steel, make it ideal for tough, large-scale welds.

FCAW-S uses flux-cored wire that generates its own shielding gas when heated, which means:

No Gas Needed: *No cylinders, regulators, or rental fees.*
Wind-Resistant: *Works well outdoors*
Portable: *Grab it, go, and weld anywhere.*
Quick Tacks: Fast, strong welds with deep penetration.

Tiny project? Precision still isn't optional.

📖 8.6 Brief Evolution of Welding (See "Welding Through the Ages: From Hammer to Laser" in Appendix G)

Welding dates back to 3000 BCE, when blacksmiths fused metals with heat and hammering. In the Middle Ages, forge welding was used for weapons and armor. Arc welding began in the 1800s after Sir Humphry Davy's arc discovery, and by the 1890s it was replacing fire-based methods. The 1900s saw the rise of major types like SMAW, gas, SAW, GMAW, GTAW, and FCAW. Since the 2000s, welding has advanced with laser, electron beam, robotic, and fiber technologies.

8.7 Shielded Metal Arc Welding (SMAW)

Also known as stick welding, SMAW uses flux-coated electrodes that generate their own shielding gas as they burn, no external tanks needed. All you need is a rudimentary welder and a box of rods.

Minimal Setup

No gas tanks, no fancy gear. My first welder was a Lincoln AC Tombstone, basic, reliable, and still going strong. Perfect for shops, garages, or field repairs. AC is different than DC welding but you get the hang of this pretty quick.

Weatherproof

SMAW handles wind and outdoor conditions better than GMAW or GTAW as they are highly sensitive to wind. It's ideal for fieldwork and remote jobs, just use the right rod.

All-Position Welding

Simple, tough, and ready for anything, SMAW is the workhorse of the welding world. It works in every position:

- 1 = All positions (flat, horizontal, vertical, overhead)
- 2 = Flat and horizontal
- 3/4 = Limited use

8.8 Commonly Used SMAW Rods

Each rod has its superpowers and weaknesses, so let's pick the one that'll help you save the day (or at least make your project awesome). There are 100s of electrode options BTW.

Electrode	Strength	Limitations
E6010	*Deep penetration, great for fieldwork and repairs*	*Difficult to strike and maintain an arc.*
E6011	*Versatile, works on dirty or rusty metal, all positions*	*Produces more spatter.*
E7018	*Strong, clean welds, minimal spatter, ideal for structural work*	*Requires moisture-free storage.*
E7014	*Easy to use, good for general fabrication, smooth beads*	*Limited penetration vs. E7018.*
E7024	*Fast welds, high deposition, great for flat joints*	*Not suitable for vertical or overhead.*

8.9 Understanding SMAW Rod Classifications

SMAW rod classifications might look like a jumble of numbers, but they pack in key info, like tensile strength, welding position compatibility, and flux coating type.

My go-to is the 1/8" E7018, a solid all-around rod. I spent two years at Niagara College mastering it in all positions, both fillet and groove welds, while training for simulated Canadian Welding Bureau (CWB) certification tests.

The '70' means 70,000 psi tensile strength, while the '18' tells you it's a low-hydrogen electrode with an iron powder coating, great for reducing spatter and slag while delivering smooth, reliable welds. It's a go-to for structural steel and heavy fabrication.

Just remember: E7018 rods are moisture-sensitive and should be stored in a rod oven to stay to code.

✐ Design for easy access to welding areas.

8.10 Welding Tips & Tricks (FCAW-S + SMAW)

Welding problems can feel like a bad romance novel, full of sparks and drama, but with the right fixes, you'll be back on track faster than you can burn a hole through your clothes.

No welding process is truly foolproof, but the most common issue I've run into is a poor grounding connection, just like with CNC plasma, a clean ground is everything.

The second most common issue? Wrong polarity. Generally, SMAW runs electrode positive (+), while FCAW-S runs electrode negative (–). Easy enough to switch on modern welders, you'll know when you have it wrong.

Challenge	Common Cause	Solution
Too much spatter	Poor settings or dirty base metal	Clean surfaces, fine-tune voltage, use proper technique.
Weak penetration	Arc too short or wrong angle	Hold correct arc length (≈ rod diameter), keep steady work angle.
Electrode sticking (SMAW)	Amperage too low or damp rods	Increase amperage slightly, keep rods dry.
Inconsistent bead size	Irregular travel speed or hand motion	Practice steady pace and consistent weave.
Welding over rust / paint	Contamination of weld pool	Always grind or wire brush before welding.
Poor visibility of weld pool	Incorrect helmet settings or lighting	Adjust shade, improve task lighting.
Electrode angle issues	Awkward body positioning	Reposition work or support your arm for better control.

8.11 Pipe vs. Tube: What's the Difference?

At first glance, pipes and tubes seem like identical twins, cylindrical and hollow, but trust me, they're not.

Understanding the difference is crucial, like knowing the difference between a hammer and a mallet. Choosing the wrong one could lead to a welding disaster, and nobody wants their project looking like a Pinterest fail.

Whether you're making a custom sculpture or working on a hydraulic system, knowing if you're dealing with a pipe or a tube is key as their sizing specs don't match.

Like many conventions in metal fabrication, such as lower gauge numbers meaning thicker material, pipe and tubing specifications can seem illogical at first.

Welding Essentials	Pipe	Tube
Material Thickness (gauge)	Thicker walls often require robust welding techniques like SMAW or GTAW	Thinner walls, precise welding, often GMAW or GTAW
Fit-Up and Alignment	Essential for clean, precise welds to prevent leaks, especially for plumbing and gas lines	Important, but may involve different joining techniques like T-joints or corner joints for structural use
Welding Positions	Challenging in vertical or overhead positions due to the round shape and hollow nature	Easier to handle in various positions due to consistent thickness, but still tricky in odd positions
Material Variety	Common materials include steel, stainless steel, aluminum, and copper, ideal for high-pressure or chemical systems	Range of materials, but often used in structural framing, requiring different filler metals and techniques for welding

📺 8.12 Tube & Pipe Welding Cost Saver

Why buy pre-cut tubing or pipe when you can cut your own and keep more cash in your pocket? I always order full-length materials and cut them myself at CNCROi.com. It's a win-win situation: it saves me money, gives me better control over precision, and leaves less scrap.

Cutting in-house also means you can tweak things mid-project and ensure a cleaner, stronger weld, no mess, no stress. All metal fabrication should be this way!

8.13 Cold Cutting (e.g., saws, waterjets)

Cold cutting, like a waterjet or carbide saws, have many advantages over abrasive methods.

Pros:
- **No HAZ:** Keeps metal relaxed, no warping, no stress.
- **Clean Cuts:** Weld-ready edges, no grinding marathons
- **Safe & Tidy:** No sparks, no dust, no fire risk
- **Ideal for thin metals:** No warping

Cons:
- **Slowpoke:** Precision takes time, don't rush art.
- **Pricey Gear:** Waterjets aren't cheap, nor carbide blades.
- **Parts Wear:** Blades and nozzles have expiration dates.

Abrasive Chop Saw
Cuts like it's late for lunch, loud, fast, and little finesse.

Pros:
- **Speed Demon:** Great for bulk cuts
- **Budget-Friendly:** Basically the fast food of cutting tools
- **Portable:** Toss it in the truck and go.

Cons:
- **Sparks Galore:** Hot metal confetti everywhere
- **Messy Business:** Loud, dusty, and a cleanup nightmare
- **Rough edges:** Needs prep, kills blades fast.

8.14 Welding Symbols (a.k.a. Weld Hieroglyphics)

Welding symbols are the secret code of fabrication, tiny sketches loaded with info: weld type, size, location, and all the specs that keep things from falling apart.

I spent two years decoding these cryptic Hieroglyphics in Mike's drafting classes. It felt more like ancient Egyptian than a course, but totally worth it. Getting them right means fewer mistakes, less grinding, and way fewer "oops" moments on the shop floor.

This book leans more into hands-on fabrication than technical notation, but learning these symbols is like giving your designs Google Translate between the office and the shop. *If you truly want to become a welder, learn it.*

Want to dig deeper? *Blueprint Reading for Welders by A.E. Bennett and Louis J. Siy* is a great start, like a Rosetta Stone for welders, minus the subscription.

I used my welding fixture table top to build the base, ensuring the ENTIRE assembly was flat, square, and parallel.

Firepits: where fabrication meets backyard bragging rights.

I was tired of flimsy store-bought fire pits, thin metal that warps, rusts, and collapses after just a few years.

So, I built my own: tougher and way more durable.

The Problem:
Most commercial pits use thin sheet metal that can't handle the heat and fail fast. I wanted something durable, and it was the perfect excuse to sharpen my plasma cutting and welding skills while in school.

The Solution:
I designed a fire pit from 3/16" (4.7 mm) hot-rolled mild steel, thick enough to take a beating. It's 3 feet wide (91 cm), 2 feet tall (61 cm), with solid 2" x 2" legs (1/8" wall), and a rugged 1.5" high grate made from 1/8" wall angle iron. Built like a tank, burns like a dream with lots of airflow.

If your table is level and flat, so should your work!

Fabrication Process

1. Plasma Cutting: Chose plasma cutting for its speed and precision.

2. Welding: For the welds, I used Self-Shielded Flux Core Arc Welding (FCAW-S) for the initial tack welds and Shield Metal Arc Welding (SMAW) for the full structural welds. A segmented and staggered approach reduced heat buildup, preventing warping from Heat-Affected Zones (HAZ).

3. Grate Construction: The removable grate, made from angle iron, added ventilation to improve combustion and made cleanup a breeze. It was *HEAVY*!

Innovative Design Features

Self-Cleaning Concept: Fully welded seams keep water from rusting it out where the metal would build-up, and ventilation holes at the bottom promote airflow, reducing moisture buildup.

Removable Grate: Easy to clean and helps the fire get going faster.

Heat Resistance: Coated with heat-resistant paint that can handle temperatures up to 2,000°F (1,093°C), though even this lasted but a few fire's worth.

The Result: This custom fire pit is built to endure the elements and frequent use. The removable grate and ventilation system make it easy to maintain and keep it burning hot.

Another One Off the Table

This fire pit project is a perfect example of how custom fabrication can solve the problems that off-the-shelf products can't.

With the right skills, tools, and a little creativity, I crafted a fire pit that's not only functional but built to last.

This project shows the beauty of hands-on learning, making it your own and solving real-world problems one weld at a time!

🏅 8.16 Practical Takeaways from the Shop

Master the basics: Fancy tools won't help if you skip fundamentals, clean materials and dry rods are key. Otherwise, expect spatter and weak welds.

Keep arc length in check: Too short or too long ruins welds. Like using the wrong tool, poor arc control means weak bonds.

Tune your settings: Wrong voltage or amperage causes spatter and weak welds. Balance your settings like tuning an instrument.

Use the right process: Each method has its place, SMAW for outdoors, GTAW for precision. Choose wisely for the best results.

✖ 8.17 Chapter 8 Quiz

Ready for a quiz? Let's see if your welding knowledge is as solid as your beads!

1. What's the main benefit of FCAW-S over GMAW?
   A) Cleaner welds
   B) No need for external shielding gas
   C) Faster welding speeds
   D) Higher quality welds

2. Which welding process is best for the great outdoors?
   A) GMAW
   B) SMAW
   C) FCAW-G
   D) GTAW

3. Which welding method needs external shielding gas?
   A) SMAW
   B) FCAW-S
   C) GMAW
   D) Both A and B

4. What's the key advantage of using SMAW for welding?
   A) It is weatherproof and works well outdoors.
   B) It offers the cleanest welds.
   C) It requires complex setups.
   D) It is ideal for cosmetic welds.

5. What does the '70' in E7018 SMAW rod mean?
   A) The amperage setting required for the rod
   B) The tensile strength of the weld in pounds
   C) The voltage needed for the weld
   D) The type of coating used

6. What's a disadvantage of using FCAW-S?
   A) It creates more spatter than GMAW welding.
   B) It is difficult to use in windy conditions.
   C) It requires external shielding gas.
   D) It is only suitable for thin materials.

7. What's the best welding process for precision welds?
   A) SMAW
   B) FCAW-S
   C) GMAW
   D) GTAW

8. What's "HAZ" in welding, and why does it matter?
   A) HAZ (Heat-Affected Zone): it impacts material strength.
   B) High Arc Zone: it controls the weld pool temperature.
   C) High Amperage Zone: it affects electrode behavior.
   D) Heat Application Zone: it's the safest welding area.

9. What's a key factor in choosing pipe vs. tube for welding?
   A) Pipes use SMAW, while tubes use GTAW for welding.
   B) Pipe is for framing, tube is for fluid transport.
   C) Tube walls are always thicker than pipe walls.
   D) Tube transports fluids, pipe frames structures.

10. What's a typical disadvantage of using SMAW?
    A) It requires external shielding gas.
    B) It isn't effective for vertical or overhead positions.
    C) Rods must be vacuum-sealed to prevent moisture.
    D) It produces a rough, uneven bead finish.

11. What does the first digit in SMAW rod classification mean?
    A) The welding position the rod can be used in
    B) The type of material the rod can weld
    C) The tensile strength of the rod
    D) The type of coating the rod has

12. What's the advantage of cold cutting over abrasive saws?
    A) Faster cutting speeds
    B) No heat-affected zone (HAZ)
    C) Lower upfront cost
    D) More portability

13. Which welding process suits thick materials?
    A) GMAW
    B) FCAW-G
    C) SMAW
    D) MCAW

14. What's a common mistake when learning welding?
   A) Using the wrong electrode size
   B) Holding the torch at the wrong angle
   C) Ignoring shielding gas needs
   D) Using too much heat

15. Why is easy access to welding areas important?
   A) It makes the welds look more professional.
   B) It improves the quality of the weld.
   C) It reduces the time spent welding.
   D) It minimizes the amount of heat needed.

16. Which welding method works best in windy conditions?
   A) SMAW
   B) GMAW
   C) FCAW-S
   D) GTAW

17. Why choose FCAW-S for fieldwork?
   A) It produces cleaner welds.
   B) It doesn't require external gas.
   C) It is best for cosmetic finishes.
   D) It works only on small parts.

18. Why is GTAW ideal for aluminum welds?
   A) It doesn't require any shielding gas.
   B) It produces a fast, rough weld.
   C) It is highly precise and clean.
   D) It is cheap and easy to set up.

19. What would you most likely use an abrasive chop saw for?
   A) High-precision cutting for aerospace parts
   B) Quick cuts for metal framing or fabrication.
   C) Cutting thin, soft materials
   D) Cutting materials with minimal heat-affected zone

20. What's true about welding pipes vs. tubes?
   A) Pipe walls are thinner than tube walls.
   B) Tube welding uses GTAW, pipe welding uses SMAW.
   C) Pipes are for structure, tubes are for fluid transport.
   D) Tube welding is less precise than pipe welding.

Using a fixture table? Welcome to the easy mode.

My old welding table 2018 vs my fixture table 2025.

✖ 8.18 Answer Key + Explanations

Ready for the breakdown? Here's the answer key, with all the weld-worthy explanations to back it up.

1B – FCAW-S uses flux-core wire that generates its own shielding gas, ideal for portable or field use without gas tanks.

2B – SMAW is reliable for outdoor and field welding without external gas.

3C – GMAW needs external shielding gas to prevent weld contamination.

4A – SMAW works well in tough outdoor conditions.

5B – "70" means 70,000 psi tensile strength, key for rod selection.

6A – FCAW-S makes more spatter than GMAW but is faster and better for field use.

7D – GTAW is precise and ideal for clean welds on delicate metals like aluminum.

8A – The HAZ can change material properties like hardness, affecting weld quality.

9A – Pipe welding often uses SMAW; tube welding usually requires GTAW for precision.

10C – E7018 rods must be kept dry to prevent cracking.

11C – The first two digits of an SMAW rod indicate tensile strength in 1,000 psi.

12B – Cold cutting keeps material hardness intact, ideal for heat-sensitive metals.

13D – MCAW gives high deposition rates, perfect for thick material.

14B – Poor torch angles cause messy welds; good technique matters.

15B – Easy access improves weld control and quality.

16C – FCAW-S is great outdoors, no gas tanks needed, wind-resistant.

17B – FCAW-S is gas-free, perfect for portable field welding.

18C – GTAW excels in clean, precise welds on sensitive materials.

19B – Chop saws are fast but create HAZ and rough edges.

20B – GTAW is preferred for precision tube welding.

12 ga 304SS 2B blank straight off the CNC plasma.

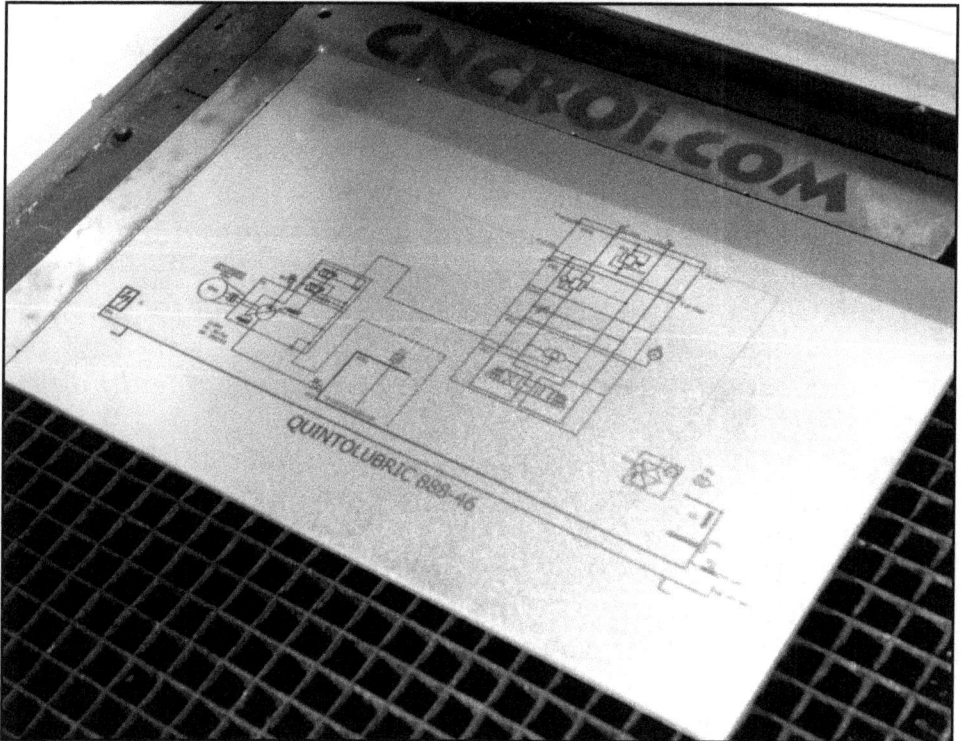

Same plaque, after clean-up and fiber laser etching in our laser.

Chapter 9
CNC Nesting Optimization Tactics
Unlock Bigger-on-the-Inside Nesting Strategies

CNC nesting is where smart design meets spatial strategy, unlocking "bigger-on-the-inside" tricks to maximize every sheet. Whether using a router, waterjet, laser, or plasma, nesting isn't just setup, it's a tactical edge.

Done right, it boosts yield, cuts waste, and controls wasted material and machine time. But it's more than clicking "auto-arrange" and hoping for the best. Real optimization means planning, cut order, part orientation, grain direction, and offcuts with future use in the shop.

Let's cover nesting tactics that stretch stock, shrink scrap, and pack more into less, like a TARDIS for toolpaths. This is the final step before production begins!

One person's scrap wood is another person's bonfire fuel.

💬 *Nesting turns digital chaos into material efficiency, one puzzle at a time.*

9.1 What's Nesting?

Nesting is the process of arranging shapes on a material sheet to maximize space. Think of it like Tetris, but with real money at stake. This process can make or break profit.

The better the nesting, the more parts you get per sheet. Since it's such a critical part of any production workflow (CNC or otherwise), I'm expanding on it here after briefly introducing it in Chapter 2.

👎 *Maximizing part count without factoring in grain, cut order, or material limits.*

9.2 CNC Nesting Applies To All CNC Systems

While lasers often steal the spotlight for their ultra-tight nesting capabilities, any sheet-based CNC system, plasma, waterjet, mill, router, or laser, benefits from it. If the machine uses a sheet, board, or plate, it can (and should) be nested. Nesting isn't process-specific; it's material-smart.

Nesting for Efficiency: Manual, Automatic & Hybrid

Nesting, the art of arranging parts to get the most out of your material, is a cornerstone of CNC efficiency. Whether you go manual, automatic, or use a hybrid approach, your choice directly impacts time, cost, and cut quality.

Yes, this chapter repeats a few points, because this stuff really matters. Better nesting means less waste, faster production, and a smoother workflow.

Method	Strengths	Limitations
Manual	*Full control, ideal for unique jobs*	*Is time-consuming and error-prone.*
Automatic	*Fast, optimized for minimal waste*	*Ignores grain, cut order, and aesthetics.*

9.3 Why Do Most Shops Go Hybrid?

Let software handle 80% of the layout, then fine-tune manually for:
- grain direction,
- material flow, and
- visual balance.

This approach blends speed with precision. Shared cut edges and efficient layouts reduce laser time, tool wear, and material waste.

Whether nesting is manual, automatic, or a mix of both, it leads to faster jobs, less waste, and better results.

9.4 Key Nesting Considerations

Nesting isn't just about cramming parts together like a rushed jigsaw, it's about cutting them right. When done properly, it leads to cleaner cuts, stronger parts, and far less swearing during assembly.

It also maintains consistent edge quality, helps prevent heat-affected warping (especially with plasma or laser), and reduces tool wear by minimizing unnecessary moves.

Factor	Why It Matters
Grain Direction	For wood/metal aesthetics or strength
Kerf Width	Especially in plasma / laser, avoid part collision or warping
Part Orientation	For logo alignment, text readability, mechanical function
Heat Distribution	In plasma / laser cutting to avoid warping or distortion
Tab Placement	Holding small parts in place during cutting
Toolpath Order	Cutting interior details before outer contour to avoid part movement.

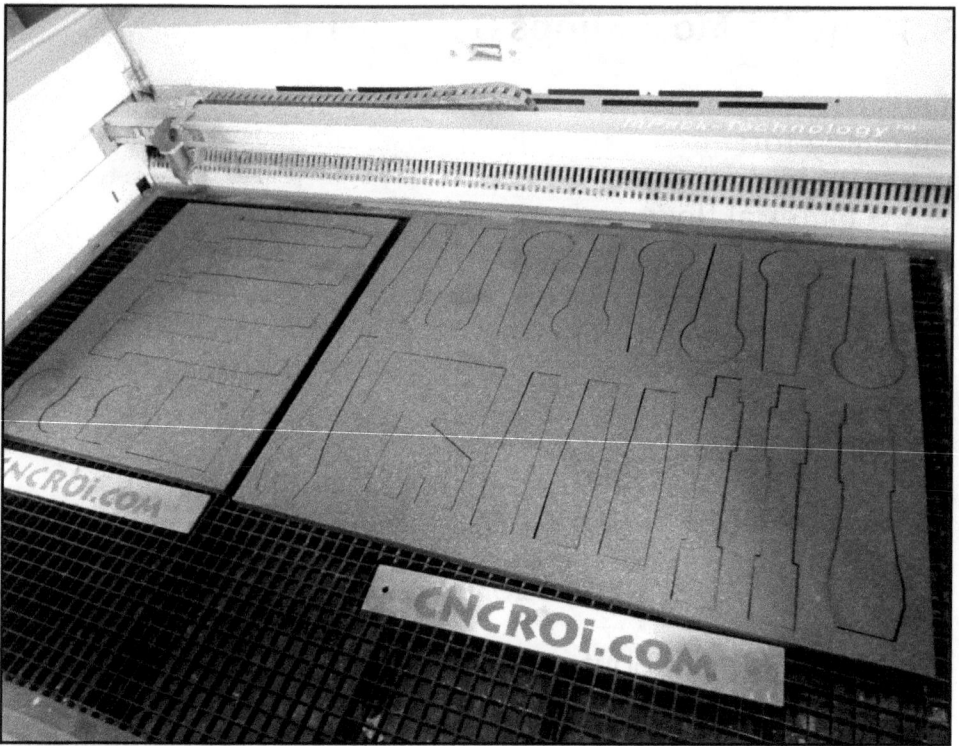

Leverage the kerf, especially with ½" foam.

9.5 Nesting in Workflow

Nesting starts long before the cut, even before you design. From the start, always ask:
- How many parts will I need?
- Can they share a material sheet?
- Is there a repeatable layout I can save?

Efficient nesting starts at the drawing table or monitor.

⌻ 9.6 Nesting = More Margin

Smart nesting is like the coupon code CNC didn't know it needed, more parts, less waste, every time.

Without it you burn through material, waste time on setups, and end up with a scrapyard's worth of offcuts with no home. With nesting you pack parts tighter, run longer without stopping, and breeze through QA with tidy layouts.

9.7 Nesting Across Materials

Nesting isn't one-size-fits-all, each material has its own quirks, like a cast of oddly-behaved roommates in a CNC sitcom. Here's how to keep everyone in line and avoid dramatic meltdowns (literally, in the case of acrylic).

Some materials have fewer issues as they get thicker, others as they get thinner. Did I mention to always test first?

Material	Challenges	Pro Tip
Acrylic	Heat buildup, melt edge	Space parts to allow cooling.
Aluminum	Prone to burring and heat distortion	Use sharp tools, allow for chip removal and cooling spacing.
Cardboard	Flammable, can shift or curl	Use minimal power and multiple hold-down points.
Foamboard / Vinyl	Lightweight, prone to shifting	Use vacuum bed or hold-downweights.
Leather	Can scorch or curl at edges.	Space cuts to allow venting and reduce discoloration.
Plywood / MDF	Grain direction, chip-out	Orient parts with grain; tabs help stability.
Polycarbonate	Melts easily, prone to edge fogging.	Use sharp tools, lower speed, and space for cooling.
Rubber	Stretchy, can deform during cutting.	Add support tabs and avoid overly narrow details.
Steel (carbon)	Heavy, can create slag or dross buildup.	Allow extra spacing to prevent fused edges.
Steel (stainless)	Warp from heat or waterjet pressure	Spread cuts evenly to distribute heat/load.

📱 9.8 Success Story
Plywood Big Wheels

Took weeks to design, cut, and build, loved every minute!

My largest project yet involved crafting a complex design from 263 pieces across 10 full sheets (4 × 8 ft or 1200 × 2400 mm), all using my trusty ShopBot Desktop, no less!

With equipment constraints, I had to divide each sheet precisely, and some parts had to flow seamlessly across multiple tiles.

Challenges and Solutions

1. Limited Equipment
My ShopBot Desktop CNC router couldn't handle full sheets, stretching the project timeline from a few days to two weeks.

Solution:
I split each sheet into manageable sections, allowing me to work within the router's limits.

2. Assembly Mistakes
I installed the pedal mechanism incorrectly, which caused movement issues.

Solution:
Disassembled and reassembled the parts in the correct orientation.

3. Material Thickness
Slight variations in plywood thickness caused issues with wheel fitment.

Solution:
I scaled up the design by 10% to compensate for the material discrepancy.

4. Steering Issues
The steering column was too heavy, making it difficult to turn.

Solution:
I reinforced the design to improve strength and balance.

5. Rolling Functionality
Excessive friction on the back wheels when under weight.

Solution:
I left this as a future improvement, planning to swap to a smoother material. Now I know why car axles aren't made out of plywood!

Results

The 30 kg project was completed using only interlocking parts and glue, no screws or nails. Despite the challenges, it was a success and proved the value of adaptability in custom fabrication.

Lessons Learned
- Adapt designs to available equipment.
- Double-check assembly for accuracy.
- Plan early for material variations.

Tiling to make large parts using a small machine.

Make parts unique for easy sorting.

Another One Off the Table

Creative problem-solving is key to overcoming obstacles in custom manufacturing. I revisited this project a few years later, this time cutting it on my Trotec Speedy 400 flexx using 1/8" (3 mm) MDF, and it fit perfectly!

Here's the bite-sized laser version, same files, just shrunk down!

🏅 9.9 Practical Takeaways from the Shop

Design with nesting in mind: Don't wait for "auto-arrange", plan like a road trip: map first, drive later.

Offcuts aren't trash: Use them for engraving, prototyping, or store them smartly for future use.

Work with material orientation: Follow the grain or stress direction, it cuts cleaner and stays flatter.

Nest smart, save more: Less waste, faster jobs, higher margins.

✖ 9.10 Chapter 9 Quiz

Ready to test your layout logic? This quiz will reveal whether you're a nesting ninja or a guess-and-go guru.

1. What's the primary goal of nesting in CNC manufacturing?
    A) To increase the complexity of designs
    B) To make the most efficient use of material
    C) To create designs that are difficult to cut
    D) To reduce the number of parts in a project

2. Which CNC systems benefit from nesting?
    A) Only laser cutters
    B) Plasma cutters
    C) Routers and waterjets only
    D) Any sheet-based CNC system

3. What's a common beginner mistake in CNC nesting?
    A) Maximizing material usage without considering fit
    B) Ignoring the cost of material
    C) Maximizing parts without minding grain or limits.
    D) Using automatic nesting tools too often

4. Which nesting method is slow but fully controlled?
    A) Automatic nesting
    B) Hybrid nesting
    C) Manual nesting
    D) No nesting

5. What's a key advantage of hybrid nesting?
    A) It's the fastest method available.
    B) It fully controls part orientation and grain direction.
    C) It avoids using CNC machines entirely.
    D) It minimizes the need for manual adjustments.

6. Which factors matter for CNC nesting in wood and metal?
    A) Grain direction
    B) Toolpath speed
    C) Kerf width
    D) Both A and C

7. In CNC nesting, what's kerf width?
    A) The thickness of the material
    B) The width of the cut made by the tool
    C) The design of the cut path
    D) The spacing between parts

8. Why is part orientation important in nesting?
    A) For fitting more parts in less space
    B) For aligning logos or text correctly
    C) To minimize heat distortion
    D) To save cutting time

9. What matters when nesting for efficiency?
    A) The toolpath order
    B) Part thickness
    C) The material's grain and cut order
    D) All of the above

10. How can smart nesting affect your bottom line?
    A) It increases waste handling costs
    B) It creates unpredictable layouts
    C) It leads to longer setup times per part
    D) It reduces material usage and increases part yield

11. What's a nesting challenge with plywood or MDF?
    A) Avoiding heat buildup
    B) Managing grain direction and chip-out
    C) Preventing part shifting due to light material
    D) Distributing heat evenly

12. What's one benefit of manual nesting?
    A) It automatically optimizes material usage.
    B) It allows fine-tuning like grain adjustment.
    C) It is the fastest method for large batches.
    D) It doesn't require any post-processing.

13. What should you watch for when nesting acrylic?
    A) The material's thickness
    B) Grain direction
    C) Heat buildup and melt edges
    D) Part orientation for aesthetics

14. When must heat distribution be considered in nesting?
   A) When cutting stainless steel with plasma or laser
   B) When working with foamboard
   C) When working with plywood
   D) When cutting acrylic

15. What challenge did Jon face using a desktop CNC?
   A) Overestimating the material thickness
   B) Not aligning parts correctly
   C) Limited cutting bed size capacity slowed the project
   D) Incorrect toolpath selection

16. How did Jon handle material thickness variation?
   A) Switched to a different material
   B) Scaled the design by 10%
   C) Ignored the variation
   D) Used a thicker plywood

17. What's the best way to hold small parts down?
   A) Gluing them to the bed of the machine
   B) Using heavier tools
   C) Adding extra material
   D) Adding tabs to the part

18. What's a key takeaway from Jon's project?
   A) Always use manual nesting for better control
   B) Avoid cutting small parts
   C) Don't change part orientation mid-project
   D) Adapt designs to equipment and material limits.

19. How does grain direction affect nesting in materials?
   A) It only affects the design aesthetic.
   B) It affects the strength and finish of the material.
   C) It doesn't matter unless the material is thin.
   D) It only affects the cutting speed.

20. Should you nest re-run parts too?
   A) **YES!**
   B) Nesting is for the weak.

✖ 9.11 Answer Key + Explanations

Check the answer key to see where your nesting instincts were spot-on, or off the mark!

1B – Nesting arranges parts to minimize waste and cutting time.

2D – It applies to all sheet-based systems, not just lasers.

3C – Focusing only on part count ignores grain direction, affecting strength and aesthetics.

4C – Manual nesting allows control but is slow and error-prone.

5B – Hybrid nesting combines speed with manual tweaks for grain and looks.

6D – Grain direction and kerf width both affect precision and results.

7B – Kerf is the cut width, critical for fit and spacing.

8B – Orientation matters for logos, text, and part function.

9D – All factors listed impact nesting efficiency and quality.

10D – Smart nesting cuts material use and boosts part yield.

11B – Plywood and MDF need grain-aware layouts to avoid chip-out.

12B – Manual nesting allows flexible design tweaks for fit and grain.

13C – Acrylic needs spacing to avoid melting and bad edges.

14A – Even heat distribution avoids warping in metal cutting.

15C – Jon's desktop CNC couldn't handle full sheets, slowing progress.

16B – Scaling the design helped adjust for material thickness and fit.

17D – Tabs hold small parts in place to prevent shifting.

18D – Designs must match machine and material capabilities.

19B – Grain direction affects strength and visual finish.

20A – Nest often, big or small, prototype or final. Practice makes perfect.

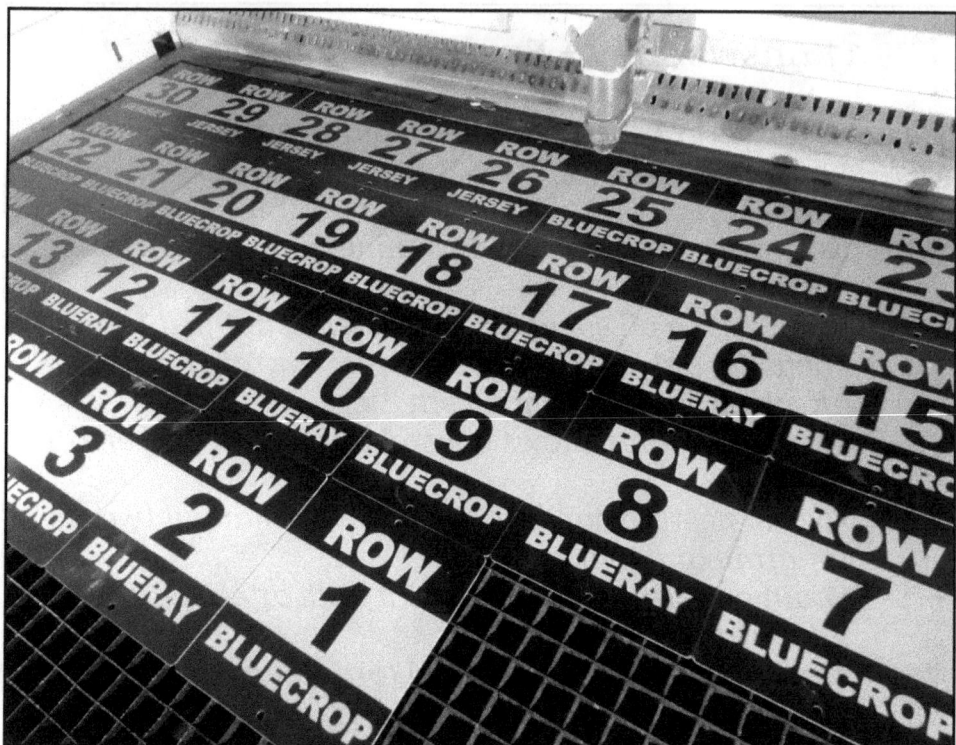

Drunk-Proof Vineyard Markers

Mid-Air Magic, Frozen in Wood

Chapter 10
Mastering the CNC Superheroes
Strengths, Weaknesses, and Synergies

Every CNC machine has a superpower, and like any good team of heroes, knowing their strengths and weaknesses is key. CNC isn't one-size-fits-all. It's more like building your own fabrication Avengers, each tool ready to save the day.

Choosing the right method is mission-critical. Use plasma where a laser belongs, and you'll regret it. Try routing O1 tool steel, hardened @ 65 HRC? That's not going to end well.

We'll cover the core four: plasma, waterjet, laser, and router. Each has strengths, flaws, and ideal use cases.

💬 *Machining is just a step before finishing.*

Every hero gets their hands dirty.

10.1 The CNC Avengers: Choosing the Right Hero

Not all CNC machines wear capes, but each has its own superpower, and its own weak spot, like the Hulk's temper or Iron Man's battery life. Choose wisely, and you'll save the day, along with tons of frustration, wasted material, and unnecessary rework.

CNC Plasma, The Hulk
Big, bold, and brutally efficient. Plasma slices through thick steel and aluminum with the grace of a wrecking ball, perfect when power and speed matter more than finesse.

Rough around the edges. Plasma excels at cutting thick metals quickly but leaves a heat-affected edge and often requires post-processing for a clean finish. It's not ideal for precise cuts, especially on thin materials.

Heavy industry hero, great for thick metals in construction, automotive, and aerospace.

CNC Waterjet, Captain America
Old-school reliable with modern precision. It uses high-pressure water and abrasives to cut nearly anything, metal, stone, glass, without breaking a sweat or generating heat.

Slow! Waterjet delivers clean, heat-free cuts but can lag on thicker materials. It also takes up significant space and generates a lot of runoff, which can be a hassle to manage.

No-heat champion, ideal for diverse materials like glass, birthday cakes, and stone.

CNC Laser, Iron Man
Cool, sleek, and surgically precise. A concentrated beam cuts through metal, plastic, and wood with minimal mess. Ideal for intricate designs, sharp edges, and that 'high-tech' vibe.

👎 *Using Default Tool Paths Without Understanding Their Impact*

High-maintenance. Lasers struggle with reflective materials like aluminum and have a limited thickness range for metals. They can also be costly to maintain due to the complex technology.

Detail king, perfect for electronics, medical gear, jewelry, and sharp aesthetics.

CNC Router, Black Panther
Fast, nimble, and versatile. From carving signs to shaping soft metals, routers do it all, wood, plastics, and composites. A true all-rounder with claws (okay, bits).

Although they use bits with endless geometries for just about anything, these consumables can get expensive fast, and things can get complex quickly. Plus, they often lack the precision and durability needed for serious metalwork.

Flexible force, loves wood, signs, and 3D work in furniture and automotive sectors.

My laser-cut Mars Rover (thanks, Angus) has never looked better!

10.2 The CNC Avengers: Ultimate Synergies

Just like the Avengers, these CNC methods shine brightest when they join forces, each bringing a unique strength to the table to deliver precision, power, and performance.

I use these synergies daily at *CNCROi.com* with awesome results. Keep in mind, these synergies aren't limited to just two machines, the more platforms you have, the greater the synergy. Each additional tool opens up new possibilities and combinations that amplify precision, and efficiency.

CNC Plasma + CNC Waterjet
Plasma is your metal-busting brute, cutting thick steel and aluminum fast and furiously. But when you need clean edges or to cut heat-sensitive materials like glass or composites, the Waterjet steps in. This combo brings both muscle and finesse, ideal for mixed-material jobs or when post-processing time needs to be minimized.

CNC Laser + CNC Router
The Laser handles intricate details, engraving, and super-fine cuts in thin materials with surgical precision. The Router complements it by shaping, pocketing, and carving complex 3D forms in wood, plastics, and soft metals. Together, they're perfect for detailed signage, prototyping, or multi-layer assemblies where both aesthetics and structure matter.

CNC Plasma + CNC Laser
Need to cut thick metal fast, then return for detailed finishes or smaller internal features? Plasma gets the bulk work done quickly, while the Laser follows up with precision, great for fabricators balancing high throughput with custom finishes or variable part geometries.

CNC Waterjet + CNC Router
The Waterjet chews through the hard stuff, stone, ceramics, hardened metals, without heat damage. Meanwhile, the Router handles the finishing work on lighter materials, custom inlays, or intricate shapes. This tag team is ideal for composite projects.

10.3 The CNC Avengers: Expanding the Team

Here's a look at additional CNC "heroes", you never know what challenges might present themselves at the shop!

CNC Bending (Falcon)
Fast, efficient, and adaptable. Much like Falcon's ability to fly and adapt to various situations, CNC bending machines can process high volumes of material with repeatable accuracy. However, while it's fast and reliable, it's not always the best fit for more intricate or highly complex shapes.

CNC EDM Wire (Doctor Strange)
Perfect for intricate, delicate shapes with high precision. But, it's slow, expensive, and struggles with thicker materials, requiring controlled conditions and specialized expertise.

CNC Lathes (Hawkeye)
Accurate for cylindrical parts, ideal for high-precision, repetitive tasks. However, it's less effective for complex shapes and requires careful setup to maintain precision.

CNC Milling Machines (Thor)
Versatile and precise, perfect for complex shapes. But, like Thor's hammer, it can be overkill for simple tasks and slower than plasma or laser cutting. Skilled operators are essential to maximize its potential.

CNC Punch Press (Ant-Man)
Small but effective, much like Ant-Man's ability to handle both big and small challenges. The CNC punch press is ideal for medium to high-volume jobs where speed is key. However, it doesn't fare well with more intricate designs or low-volume runs, where the cost of tooling could outweigh the benefit.

CNC Stamping (Black Widow)
Fast and efficient for mass production, especially in metal sheet work. However, it's limited to high-volume, identical parts and requires costly custom dies for each design.

⬜ 10.4 Success Story
Christmas Ornaments

Fiber cut 16 ga aluminum, ready for GTAW.

I took on the challenge of creating custom aluminum Christmas ornaments in various sizes for a client. This project involved a lot of prototyping, testing, and adjusting to ensure the ornaments were both beautiful and functional.

Process Breakdown

1. Prototyping and Design Adjustments

Before diving into mass production, I made small prototype samples to see how the designs worked in real life. This gave the client the opportunity to make any last-minute tweaks.

Solution: Testing the prototypes allowed the client to evaluate the assembly process and make final design decisions without the stress of mass production mistakes.

2. Laser Cutting Aluminum

Aluminum, while light and strong, can be tricky to cut, especially when you consider the issues with melted residue (draw) and material sticking to tools.

Solution: I opted for a laser cutter. Unlike CNC milling or routing, the laser produces minimal smoke and there is no binding, leading to cleaner cuts and less hassle.

3. Material Handling
Handling thin 16-gauge aluminum can be a bit of a balancing act, use too much power, you create a lot of dross on the underside, not enough, and you don't have a clean cut that's easy to remove from the sheet.

Solution: Lots of testing to get the settings just right!!!

Materials & Equipment

Material: 16-gauge aluminum, lightweight, cool to the touch, and stable during the laser cutting process.

Equipment: Laser cutter, chosen for its precision and minimal residue, which was important when compared to alternatives like waterjet or CNC routers.

CNC plasma was just too heavy of a lift on this, details too small and gauge too thin with intricate details.

Production and Results

I produced four sizes to meet the client's needs, giving them physical samples to evaluate and decide which worked best for their product line and display scenarios.

Once assembled using GTAW, the client had a clear roadmap for mass production.

The prototyping phase removed guesswork, reduced assembly issues, and confirmed the final product's structural integrity and visual appeal.

Kilowatt-class fiber lasers are fast, fun, and cost a small fortune.

Fiber lasers slice the skinny stuff, plasma wrestles the thick stuff.

Things I Had to Learn the Hard Way

Prototyping is essential

Fine-tuning designs before committing to full production can save time and money in the long run.

Choosing the right machine matters

Matching the material to the right cutting method is crucial for both efficiency and quality.

Another One Off the Table

My work on the custom aluminum ornaments demonstrates the power of prototyping and the need for a strategic tool selection in custom manufacturing. The key takeaway? Don't just cut corners, cut smart.

🎖 10.5 Practical Takeaways from the Shop

Specialties matter: Like the Avengers, each CNC method has a distinct role. Plasma brings brute force, waterjet delivers heat-free precision, laser offers fast accuracy, and routers provide unmatched versatility. Choosing the right one makes all the difference.

Know the trade-offs: Plasma can leave rough edges, waterjet is slower and messier, lasers have material limitations, and routers aren't made for heavy metals. Knowing these helps you avoid costly missteps.

Team-ups win: Combining methods, like a superhero crossover, often leads to better results. Plasma and waterjet blend speed with precision; laser and router mix detail with 3D shaping.

Respect your machines: CNC tools are your fabrication Avengers, treat them right, and they'll save the day. Push them too hard, and you might unleash a project-wrecking villain.

❌ 10.6 Chapter 10 Quiz

Take this CNC Superhero Quiz and find out, are you the real deal or just wearing the cape?

1. What's the main advantage of CNC plasma cutting?
   - A) It is highly precise for intricate cuts.
   - B) It excels at cutting thick metals quickly.
   - C) It cuts delicate materials without heat distortion.
   - D) It is best for detailed woodwork.

2. Which CNC method is precise with minimal heat distortion?
   - A) CNC Plasma
   - B) CNC Waterjet
   - C) CNC Laser
   - D) CNC Router

3. Which material is CNC waterjet cutting ideal for?
   - A) Thin metals like steel and aluminum
   - B) Stone, glass, and food
   - C) Soft metals like copper and brass
   - D) Wood and plastics

4. Which CNC method excels at engraving and 3D shaping?
   - A) CNC Plasma
   - B) CNC Waterjet
   - C) CNC Laser
   - D) CNC Router

5. What's a major drawback of CNC plasma cutting?
   - A) It leaves rough edges and heat-affected areas.
   - B) It is slower than CNC laser cutting.
   - C) It cannot cut metals thicker than 1 inch.
   - D) It requires special materials for cutting.

6. What material is difficult for CNC laser cutting to handle?
   - A) Wood
   - B) Plastics
   - C) Highly reflective metals like brass and copper
   - D) Glass

7. What's a key advantage of CNC waterjet cutting?
    A) Speed and brute force
    B) Ability to cut through reflective metals
    C) No heat distortion during cutting
    D) Cutting thicker metals faster than plasma

8. What's a limitation of CNC routers?
    A) They struggle with cutting soft materials.
    B) They are not ideal for heavy-duty metalwork.
    C) They can't cut 3D shapes.
    D) They are too slow for large projects.

9. Which CNC metal combo gives speed and precision?
    A) CNC Plasma + CNC Waterjet
    B) CNC Plasma + CNC Laser
    C) CNC Waterjet + CNC Router
    D) CNC Laser + CNC Router

10. What's a major challenge with CNC waterjet cutting?
    A) It leaves rough edges on materials.
    B) It can produce a lot of water runoff.
    C) It cannot cut through soft metals.
    D) It is the fastest CNC method.

11. Best CNC method for fine cuts in thin metal?
    A) CNC Plasma
    B) CNC Waterjet
    C) CNC Laser
    D) CNC Router

12. What does a CNC router excel at cutting?
    A) Heavy metals
    B) Wood, plastics, and soft metals
    C) Stone and glass
    D) High-speed cuts on thick materials

13. Which CNC method cuts thick metal fast?
    A) CNC Laser   B) CNC Waterjet
    C) CNC Plasma   D) CNC Router

14. What's the main CNC laser limit when cutting metal?
   A) It is too slow for large projects.
   B) It leaves behind a rough, heat-affected edge.
   C) It is only suitable for plastics and wood.
   D) It cannot handle thick metals efficiently.

15. Which CNC method cuts frozen vegetables?
   A) CNC Plasma
   B) CNC Router
   C) CNC Laser
   D) CNC Waterjet

16. What's the benefit of CNC plasma in heavy industry?
   A) Cuts thick metal fast and powerful
   B) High precision for small parts
   C) Low energy consumption
   D) Avoids heat distortion during cutting

17. What material is CNC router cutting primarily used for?
   A) Stone and glass
   B) Thin metals
   C) Wood, plastics, and soft metals
   D) Frozen food products

18. What's the key to choosing a CNC cutting method?
   A) The speed of cutting only
   B) The precision needed for the cut
   C) The cost of the machine alone
   D) Whether the machine uses heat or water

19. Why combine CNC methods for better results?
   A) To balance speed and precision
   B) To minimize the time spent on maintenance
   C) To reduce the cost of the materials used
   D) To avoid having to use different machines

20. What's the key lesson from the ornament case study?
   A) Prototyping is unnecessary if the design is simple.
   B) Always use the same cutting method for all jobs.
   C) Choosing the right CNC method is crucial.
   D) No cutting changes needed after prototyping.

Fiber cut, etched, paint-filled, yet still ugly.

When you walk up to opportunities door, don't knock it... KICK THAT BITCH IN, smile and introduce yourself.

~Dwayne Johnson

The Rock, CO_2 laser engraved on a rock, forever.

Prototype: Fail until the universe gives in.

Fail you will. Learn you must. Master you shall become.

✖ 10.7 Answer Key + Explanations

Here's your answer key, now with origin stories!

1B – CNC plasma cuts thick metals like steel and aluminum quickly, ideal for heavy-duty jobs.

2C – CNC laser offers high precision and clean cuts with minimal heat distortion, perfect for intricate work.

3B – CNC waterjet handles tough materials like stone, glass, and even food, without heat distortion.

4D – CNC routers excel at engraving, detail work, and 3D shaping, making them great for complex designs.

5A – Plasma cutting creates rough edges and a heat-affected zone, requiring cleanup.

6C – Lasers struggle with reflective metals like brass and copper, needing adjustments.

7C – Waterjets cut without heat, ideal for heat-sensitive materials.

8B – Routers are great for wood, plastics, and soft metals but not thick or hard metals.

9A – Plasma + waterjet combines speed for thick cuts and precision for delicate materials.

10B – A key challenge with waterjets is managing water runoff and cleanup.

11C – Lasers are best for small, precise cuts in thin materials.

12B – Routers are versatile, great for wood, plastic, soft metals, engraving, and 3D cuts.

13C – Plasma is the go-to for fast, thick metal cutting.

14D – Lasers struggle with thick metals compared to plasma.

15D – Waterjet is perfect for delicate cuts like frozen food, no heat, no damage.

16A – Plasma suits industries like automotive and aerospace for thick metal cutting.

17C – Routers excel in soft materials and 3D subtractive work.

18B – Choosing the right CNC method depends on the level of precision required.

19A – Combining methods like plasma + waterjet or laser + router boosts both speed and precision.

20C – Choosing the right CNC tool, like laser for aluminum, improves quality and reduces cleanup.

Your CNC runs on G-code; some clients run on delusion.

Chapter 11
CNC Robotics and Automation
Streamline Tasks While You Sleep

Automation isn't just for mass production anymore, it's transforming custom CNC work. Robotic arms now weld unique fixtures, while smart conveyors manage mixed materials with precision and ease.

These tools enhance efficiency, consistency, and scalability, even for small shops and short-run projects. With the right setup, your CNC machines can keep working long after you've clocked out.

We'll explore how automation systems, from robotic arms to intelligent add-ons, integrate with CNC platforms. Even small operations can start simple, scale smart, and stay agile while boosting output and reducing manual labor. Always remember, CNCs follow even bad instructions blindly.

A perfect lights-out CNC shop, closer than you think!

💬 *Automation executes your intent, not corrects your oversight.*

11.1 Robotics in CNC: Beyond Welding

When most people think of robots in fabrication, they picture welding arms, and for good reason. But that's just the beginning. Enter the cobot: a collaborative robot designed to work safely with humans, not behind a cage.

Robotic Application	CNC Process	Use Case
6-axis arm + fiber laser	*Laser cutting / welding*	*Cutting tubing, engraving curved surfaces, precision joints*
5-axis Robotic router head	*CNC routing*	*3D surface carving (at an angle), ergonomic shaping, deep contoured cuts*
Waterjet + robotic control	*CNC waterjet*	*Sculptural stone, intricate curves in heavy materials*
Cobot handling system	*Any*	*Material loading/unloading, repositioning, post-processing*

I hope this opens your eyes to how you can absolutely mix and match multi-axis systems with robotic tech, like a high-stakes LEGO set, building ultra-specific workflows that no sane human should be anywhere near.

The more "lights-out" your operation becomes (machines running the night shift without much oversight), the more precise, repeatable, and productive your process gets, all without anyone risking their eyebrows in the danger zone.

This is where CNC is headed, and it's moving fast.

👎 *Automating a Broken Process*

11.2 What's a Robotic Arm?

A robotic arm mimics your own, using joints, links, and interchangeable tools like grippers or lasers to handle precise, repetitive tasks with ease.

Degrees of Freedom (DOF): More axes = more complex motion and flexibility.
End Effectors: Interchangeable tools mounted at the arm's tip, from CNC heads to suction cups.
Automation: Designed for consistency, repeatability, and fatigue-free performance.
Types: *Articulated:* Flexible, multi-jointed, great for complex tasks. *SCARA:* Fast, horizontal movement, ideal for assembly. *Delta:* Ultra-fast and precise, three-arm setup for light payloads. *Cartesian:* Moves in straight lines (X, Y, Z), simple and highly accurate.

11.3 When Does Automation Make Sense?

Automation isn't about replacing humans, it's about upgrading capabilities. The best systems don't steal jobs; they give people superpowers.

Automation	Best For	Why It Matters
Robotic Arms	Repetitive / hazardous tasks	Increases safety, frees up skilled labor.
Automated Tool Changers	Multi-tool jobs	Speeds up transitions, reduces risk of manual errors.
Material Feed Systems	High-mix / high-volume setups	Keeps machines working, not waiting.
Pallet Loaders / Unloaders	Sheet / panel handling	Reduces downtime, boosts consistency.
Vision / Inspection Tools	QC and part alignment	Enables automated validation and adaptive workflows.

⬇ 11.4 My Shop's CNC Router Evolution

I'm incredibly grateful to ShopBot Tools for inventing the ShopBot Desktop, it truly changed my life!

Today, I have both the original ShopBot Desktop and a currently a Thermwood MultiPurpose M42-55DT in my shop.

Both machines deliver industrial-grade precision and top-tier output, but the key difference is production capacity. It's like comparing a chisel to a jackhammer, both can break through concrete, but one does it faster and with far less effort.

It wasn't a straightforward transition, though. Between routers, I worked with a ShopBot PRS Alpha and then an ancient Thermwood CS43, each machine pushing me further toward the high-output workflow I rely on today.

Feature	ShopBot Desktop	Thermwood M42-55DT
Target Audience	*Hobbyists, small businesses, makerspaces*	*Industrial manufacturers, cabinet shops, high-volume production*
Build Quality	*Compact, desktop-sized*	*Heavy-duty, industrial-grade*
Table Size	*24" x 18"*	*Two 5 ft square tables*
Overall Footprint	*39" x 32" x 30" (run off a laptop and a shopvac)*	*Very cozy fit in a 2-car garage with added control box, vacuum pumps & dust systems*
Tools Changes	*Manual (ATC available)*	*ATC incorporating 19 tooling options*
Feed	*120 V AC*	*3-phase*
Cost	*You prefer a motorcycle?*	*You prefer a house?*
Weight	*70 lbs (31.75 kg)*	*6,000 lbs (2,722 kg) before the bells and whistles*

11.5 Lights-Out Manufacturing

In today's CNC environment, lights-out manufacturing is the holy grail of productivity.

This model allows CNC machines and robotic systems to run unattended, overnight or all weekend, with zero human input. That means:

- CNC routers cutting while you sleep.
- Lasers etching detailed parts by moonlight.
- Robots switching tools, loading material, and stacking finished parts, all in the dark.

For custom fabrication, lights-out means:

- Higher machine utilization.
- Faster turnaround for short-run jobs.
- Less operator fatigue.
- Greater consistency across multiple setups.

When to Automate?

When it comes to running, or working towards a lights-out manufacturing facility, don't automate the exception. Automate the routine when it:

- Solves a repeatable bottleneck.
- Has a clear ROI (more parts/hour, less labor).
- Scales with your shop (modular beats monolithic).
- Reduces downtime, not decision-making flexibility.

Lights-out automation isn't about replacing people, it's about giving them more control over their time. Let machines handle the late-night grind while you rest, recharge, or focus on higher-value work.

But don't pull a Simon! (The only boss I ever really had.) He went home proud, thinking his lasers would quietly print money through the night, only to return and find a smoking crater where Evright used to be. *He's the Supreme Commander you'll see in many of my videos BTW.*

11.6 Comparing Old vs. New Automation Models

Old automation cranked out thousands of the same part, no questions asked. New automation? It's nimble, handles change like a champ, and doesn't flinch when you shift gears mid-batch. Let's break it down.

Factor	Old Automation	New Automation
Batch Size	Large (e.g., 10,000 parts)	Small (e.g., 10–100 parts)
Setup Changes	Minimal (1x setup)	Frequent (every batch)
Efficiency	High (one product)	Moderate (more changeovers)
Flexibility	Low	High
Ideal Use	Mass production	Custom or high-mix production

11.7 Modern Tools for Flexible Automation

Today's gear flexes, adapts, and hustles like a barista at morning rush. Meet the dream team making modern shops faster, smarter, and way less cranky.

Tool	How It Helps
Quick-change fixturing	Speeds job swaps, ideal for high-mix shops.
Vision systems	ID parts and align robot paths based on actual size/orientation.
Cobots with teach mode	Easy to train, no coding needed, ideal for small shop floors.
Modular conveyors	Reconfigurable for evolving workflows.
Real-time CNC feedback loops	Adjusts settings on-the-fly for better cut quality and efficiency.

11.8 Emerging Technologies in CNC

CNC just leveled up, today's tools cut faster, think smarter, and weld cooler, literally.

Tech	How It's Used in CNC
Magnetic levitation rails	Frictionless motion = hyper-precise cuts on lasers / waterjets.
AI path optimization	Reduces cut time and tool wear through smart toolpath predictions.
Real-time feedback loops	Dynamically adjusts based on material response.
Dual-arm cobots	Two "hands" for complex assembly or part manipulation.
Laser welding (auto-feed)	Joins metals with minimal heat distortion with absolute precision.

Future CNCs may be unrecognizable, and severely bend physics.

🏅 11.9 Practical Takeaways from the Shop

Start Dumb, Win Smart: Begin with tasks so dull even your goldfish wouldn't blink. If a human zones out doing it, a robot will shine. Save the artsy, fussy work for later, or for Bob, who still swears "machines will never replace craftsmanship."

Robots Aren't Psychic: Just because it's shiny with blinking lights doesn't mean it reads minds. Robots need step-by-step instructions, like interns, but with fewer snacks and more sparks if you mess up.

Plan Before You Power On: Giving a cobot a job without a clear task is like handing a toddler a chainsaw, technically doable, but expect chaos. Plan first, automate second, breathe third.

Buy for the Future, Not the Flash: Don't blow your budget on tech that's obsolete by next Tuesday. Think long-term. Great systems scale, unlike that espresso machine that tapped out after one all-nighter with your laser.

Smart automation starts simple, follows clear instructions, plans ahead, adapts as you grow, and invests for the long haul, not the hype or the latest shiny thing.

❌ 11.10 Chapter 11 Quiz

Let's test if you've been programming like a pro, or just hitting the big red button and hoping for the best!

1. What's the main job of a cobot in custom manufacturing?
   - A) To work autonomously without any human input
   - B) To work safely alongside humans
   - C) To replace human labor entirely
   - D) To perform welding tasks

2. Which CNC process benefits from robotic arms?
   - A) Laser cutting
   - B) CNC waterjet
   - C) CNC routing
   - D) All of the above

3. What's the main benefit of CNC automated tool changers?
    A) Increased production speed and reduced errors
    B) Ability to work without human intervention
    C) Reduced power consumption
    D) Fewer tools needed per job

4. What does DOF mean for a robotic arm?
    A) Number of tools the arm can hold
    B) Number of axes the arm can move along
    C) Types of materials the robot can handle
    D) Speed at which the robot operates

5. Which application benefits most from robotic arms?
    A) Handling hazardous materials
    B) High-volume batch production
    C) Repetitive or precision tasks
    D) Quality control in custom jobs

6. How does "lights-out manufacturing" benefit shops?
    A) Machines run unattended after hours.
    B) Automates the entire workflow.
    C) Reduces the need for robots.
    D) Replaces all CNC machines.

7. How do old and new automation models differ?
    A) Old was flexible; new is rigid.
    B) New focuses on mass production; old on high-mix.
    C) Old suited big batches; new handles fast changes.
    D) Old and new automation are mostly the same.

8. What's the purpose of vision systems in robotics?
    A) Inspect materials before cutting.
    B) Spot parts and guide robot motion.
    C) Cut materials without human input.
    D) Lower robot energy use.

9. What's the most common mistake when adding robots?
    A) Not clearly defining the task before automating
    B) Using robots where they don't fit.
    C) Skipping human input entirely
    D) All of the above

10. What's the main advantage of modular conveyors?
    A) They handle different materials automatically.
    B) They're faster than traditional conveyors.
    C) They reconfigure easily for changing workflows.
    D) They make robots unnecessary.

11. Why is it smart to start small with shop automation?
    A) Small systems always cost less.
    B) You can test and grow as needed.
    C) It cuts down on team training.
    D) It guarantees faster production.

12. Which tasks should be automated first in a shop?
    A) Time-consuming, repetitive jobs
    B) Tasks needing creative thinking
    C) Tasks that rely on manual finesse
    D) Complex, multi-step assemblies

13. What's the job of a robotic arm's "end effector"?
    A) Controls the arm's overall motion.
    B) Attaches tools like grippers for tasks.
    C) Programs the robot's instructions.
    D) Monitors performance and output.

14. What does "scalability" mean in automation?
    A) Increase speed without losing quality.
    B) Grow your system with production demand.
    C) Use robots only for small batches.
    D) Automate every task in your shop.

15. What sets modern CNC automation apart from old?
    A) It's built for mass production.
    B) It favors high-volume, low-mix work.
    C) It's flexible for custom, changing jobs.
    D) It needs fewer people to run.

16. Why use dual-arm cobots in manufacturing?
    A) Double output on repetitive jobs.
    B) Use fewer robots overall.
    C) Speed up basic material handling.
    D) Tackle complex tasks with two "hands."

17. Which emerging CNC automation tech is mentioned?
    A) AI-driven path optimization
    B) Wireless robot power supply
    C) Self-programming robots
    D) CNC machines with zero downtime

18. What's the benefit of real-time CNC feedback loops?
    A) Let robots run with zero human input
    B) Auto-adjust cutting for better results
    C) Replace vision systems entirely
    D) Lower automation costs

19. What's the best use of robotic arms in manufacturing?
    A) Mass-producing identical parts
    B) Complex builds and repeat tasks
    C) Cheap jobs that don't need precision
    D) Tasks that rely on constant human help

20. Why train your team to use shop robotics?
    A) Robots are only as smart as their users.
    B) Robots don't need any human input.
    C) Robots replace workers completely.
    D) Trained teams avoid costly robot repairs.

Laser etching a pan, branding breakfast from the inside out.

Eventually, robots will fix themselves too.

✖ 11.11 Answer Key + Explanations

Here's the answer key, now with automation insights, so you can pretend you meant to do it right all along!

1B –   Cobots are built to safely work alongside humans, unlike traditional robots that need physical barriers.

2D –   Robotic arms boost precision and efficiency across CNC processes like laser cutting, waterjet, and routing.

3A –   Automated tool changers reduce downtime and errors by streamlining tool swaps mid-job.

4B –   "Degrees of Freedom" refers to how many directions a robotic arm can move, essential for complex tasks.

5C –   Robotic arms thrive on repetitive, precision-based tasks like assembly, welding, and material handling.

6A –   Lights-out manufacturing means machines run unattended, even overnight, for non-stop productivity.

7C –   Old automation favored big batches. Modern automation? Flexible and perfect for fast-changing custom work.

8B –   Vision systems help robots see parts, align accurately, and adapt in real time. Basically, robot eyes.

9D –   The biggest mistakes? Vague tasks, ignoring your team, and making things more complex than they need to be.

10C – Modular conveyors reconfigure easily, making them ideal for evolving shop layouts and workflow changes.

11B – Starting small lets you test and refine automation before scaling, smarter than going all-in blind.

12A – Repetitive tasks are perfect to automate first, clear bottlenecks, free up humans, and boost efficiency.

13B – The end effector is the business end of the robot arm. It grips, welds, cuts, and does the dirty work.

14B – Scalability means your automation setup can grow as your production needs do, no full reboots required.

15C – Modern CNC automation handles change like a champ, perfect for high-mix, low-volume production.

16D – Dual-arm cobots perform complex tasks with two "hands", think assemblies or careful part placement.

17A – AI-driven path optimization helps cut smarter, reduce wear, and adapt toolpaths in real time.

18B – Real-time CNC feedback lets systems auto-adjust mid-cut for better quality and consistency.

19B – Robotic arms shine in repetitive, precise tasks, not in roles needing constant babysitting.

20A – Robots only follow what they're told. Training your team ensures those instructions don't suck.

The future of CNC: machines talking, we just nod.

Chapter 12
From Prototyping to Production
Turn Test Builds Into Scalable Systems

Prototyping is where you learn. Production is where you pretend you knew what you were doing all along. In custom manufacturing, you're not cranking out 10,000 widgets, you're making one thing 10,000 slightly different ways.

Maybe it's a one-off spaceship-shaped enclosure for a Lego jig, or just a small batch of "how the heck do we hold this?" parts. Either way, it's all about turning an idea into a real thing without crying in front of the CNC.

Here I'll walk you through the glorious chaos from napkin sketch to final delivery, showing how every tool in the shop, from CNC routers to fiber lasers to robotic arms that look like they're about to high-five you, can help you prototype fast and scale smart.

🗩 *A prototype is just a fancy way of saying "let's see what breaks first."*

12.1 What Makes a Prototype?

A prototype isn't just a rough draft, it's a physical question mark. It asks: "Does this work?" "Can we build it?" "Will it hold up?" "What fails first?"

Each version solves a piece of the puzzle, and every answer brings you closer to reliable, repeatable production.

Prototype Type	Purpose
Conceptual	*Visual proof-of-concept or idea validation*
Functional	*Tests fit, function, and interaction*
Pre-Production	*Final version before full run (with tweaks)*

12.2 The Prototyping Process

The prototyping journey can be as chaotic as it is creative, full of unexpected twists, feedback loops, and flashes of brilliance.

Let's walk through a typical job, from the first spark of an idea to a repeatable output ready for production.

1. The Spark
The client brings an idea, often rough around the edges, sometimes just a scribble on a napkin. It could be a concept, a vision, or a problem waiting to be solved.

2. The Blueprint
The concept begins to take shape as a rough design in CAD or vector software. This is where the first tangible version of the idea comes to life in a digital space.

👎 *Automating a Broken Process*

First run of prototype wooden banks, straight to the firepit.

3. Material Selection
Choosing the right material isn't just about looks, it's about balancing form, function, and cost. Will it hold up under stress? Does it deliver the desired finish? What's the bottom line?

4. Rapid Prototyping
With speed in mind, the first prototype is crafted using the quickest tools available, CNC or otherwise. This brings the design to life in the real world, fast.

5. The Test Run
Now the fun begins. Testing the prototype in its intended environment provides valuable feedback, what works, what doesn't, and what needs tweaking.

6. Refining the Design
Armed with real-world data, adjustments are made, and a second prototype is built, this time with more precision to finalize the design. This can take several rounds.

7. Client Approval
Once changes are integrated, the design is presented for approval. If everything checks out, it's time to prep for the next step: production.

8. Optimizing Efficiency
Now it's time to optimize, nesting and layout adjustments are made to maximize material usage and machine time. Efficiency is key, especially heading into production.

9. Establishing a Workflow
With the design approved, a repeatable production workflow is set up, possibly with automation added to streamline future runs and ensure consistency.

This process might look like a straight line, but in reality, it's more like a squirrel on espresso, full of zigzags, backtracking, and unexpected leaps.

It's a chaotic little dance that never seems to get old and always has new lessons to teach you.

12.3 Lessons from Scrap Heap University

Prototypes are your product's way of saying, "Not quite there yet!" Whether it's CNC plasma, waterjet, laser, or router, this is where ideas stumble before they walk.

Every hiccup or warp is a hidden lesson, not a failure. Spot the issues now, before full production turns into full panic. It's easy to fix one mistake, not hundreds or thousands.

1. Hole Misalignment: It's Not You, It's the CAD

The Issue: Holes don't line up, fasteners refuse to go in, and your "precision" looks more like a guessing game.
What to Do: Triple-check dimensions in your CAD file. When in doubt, use your CNC laser or router to do a test cut in cardboard or MDF before committing to the real deal.

If it doesn't align in a prototype, it definitely won't in production. CAD forgiveness isn't a real thing.

2. Material Warping Under Heat: The Wavy Drama Queen

The Issue: Cutting thin metals or plastics with heat-based tools like CNC lasers or plasma results in parts that bend like warm spaghetti.
What to Do: Use a CNC waterjet for heat-sensitive materials. If using a laser, reduce power and increase speed, or cut in stages. Add hold-down tabs if parts lift mid-cut.

Know your material's limits before it does a dramatic faint.

3. Assembly Awkwardness: The IKEA Curse

The Issue: Parts kinda-sorta fit… if you hold them just right, pray, and have a mallet nearby.
What to Do: Add alignment features, tabs, notches, or guides, and use test fits. CNC routers excel here thanks to their depth control.

If your assembly needs brute force, it's probably trying to tell you something. Like "fix me."

212

4. Markings Too Light: The Vanishing Label Trick

The Issue: Your engraved logo looks more like a faint suggestion than a bold statement.
What to Do: Adjust laser power/speed settings, add multiple passes, or use contrast-boosting techniques like paint fills or backers.

Visibility matters. If your branding can't be seen, it may as well not exist.

5. Finish Not Durable: Shiny Today, Sad Tomorrow

The Issue: Your prototype looks great... until someone touches it and it scratches like a lottery ticket.
What to Do: Test different coatings, powder coat, anodize, clear coat, etc. Cut identical parts and test finishes under real-use conditions.

Test finishes like you test first dates, see how they handle stress, heat, and repeated handling.

6. Tabs Too Weak: Snap, Crackle, Oops

The Issue: Alignment tabs or structural elements break during assembly like dry twigs.
What to Do: Beef up your tab design, make them longer, wider, or double them up. Routers and lasers can handle intricate tab geometry, so take advantage.

A strong prototype survives more than admiration.

7. Edge Quality Fails: When Rough is Too Rugged

The Issue: Cuts come out jagged, chipped, or rough, especially when routing composites or plasma cutting thicker metals.
What to Do: Slow your feed rate or use multiple shallow passes for cleaner edges. Switch to a sharper bit or a smaller kerf nozzle.

Your product's edges matter, unless you're going for "rustic."

8. Unexpected Fit Issues from Kerf: The Invisible Thief

The Issue: CAD looks perfect, but cutting removes material, and parts don't fit.
What to Do: Account for kerf (the material width removed by your tool) in the design. CNC plasma and waterjet users, pay close attention here.

Kerf is like taxes, you can't ignore it, and it always takes more than you expect.

Prototypes: The Smart Way to Fail Forward

Every issue you catch in a prototype is one you won't discover after spending thousands on final production. Whether you're wielding a CNC router, plasma torch, laser beam, or high-pressure waterjet, prototypes are your best chance to learn, tweak, and improve.

Build smart. Fail small. Cut better. And always listen when your prototype groans, it's probably trying to save your future self a lot of grief.

1/4" (6 mm) plywood, perfect model, tons of parts.

🖥 12.4 Prototype to Production Workflow

Clients aren't paying for your machines, they're investing in your expertise, judgment, and ability to solve problems efficiently. That's why a smooth, well-structured transition from concept to production matters. It saves everyone time, money, and stress. Here's how to make it count:

Billable Prototyping
Always charge for early development work. It's not just trial and error, it's research and development. Clients are funding the path to a better final product, and your time is part of that value.

Material Testing Phase
Use this stage to dial in settings, confirm tolerances, and explore material behavior. It's not waste, it's insurance against costly mistakes. The more you learn now, the fewer surprises later.

Client Approval Checkpoints
Build in stages for client sign-off. This keeps expectations aligned and prevents last-minute changes that can derail production or require expensive rework.

Reusing Fixturing and Templates
Design with reuse in mind. Custom jigs, fixtures, and files that can be used across batches or similar projects drastically reduce future setup time and boost consistency.

Workflow Documentation
For repeat clients, take time to document your setup, tooling, and machine parameters. This turns one-off jobs into efficient reorders and builds trust through consistency and speed.

In short, clients hire you for your brain, not your button-pushing. Strategic planning on your end translates to confidence and clarity on theirs. They're not paying for a machine to turn on, they're investing in your judgment, experience, and knack for simplifying complexity.

12.5 Nested for Speed

Just because you nested a few parts for a small run doesn't mean that same layout works when you're scaling up. True nesting power kicks in with volume, it's about optimizing across entire sheets, not just one at a time.

I learned that firsthand with my Plywood Big Wheels project. More parts, smarter nests, fewer headaches.

12.6 Scaling Work: Production ≠ Mass Production

In custom manufacturing, scaling isn't about mass production, it's about repeatability. Whether you're laser-etching, plasma-cutting, routing, or waterjetting, the goal is making five or five hundred parts that don't suck, every time. One good part is easy; consistent quality is the real challenge. I can't emphasize this enough.

Scaling smart means refining designs, setting up jigs, and locking in machine settings for predictable results. It's about anticipating material quirks, tool wear, and human error before they cost you.

You're not building a factory, you're building consistency. Custom doesn't have to mean chaos.

Scaling Element	Example in Custom Manufacturing
File prep	Clean, parameterized CAD files for minor changes
Tool path optimization	Nesting, lead-ins/outs adjusted for speed and quality
Jig / fixture design	Reusable setups reduce alignment time
Repeatable finishes	Identical surface prep and coatings
Process standardization	Material handling, post-processing, and QC

📱 12.7 UnSuccess Story
The Drais BMT Journey

Supersized version using 12 ga 304 stainless steel 2B.

The Drais BMT (Biking Multi-Tool), created by me and was designed to be a compact, versatile tool that could ride along on a keychain, ready for everyday bike rides. With a focus on durability, usability, and affordability, the tool evolved from concept to production through thoughtful design choices and precision manufacturing techniques.

Process Breakdown

1. Idea to Function

My goal was to create a small yet functional multi-tool specific to bikes, portable enough for a keychain but tough enough for real-world use.

I emphasized multi-functionality without bulk, so you'd be able to carry it around without thinking.

2. Material Selection

Durability and affordability were key. Instead of premium-priced materials like titanium, I chose 304 stainless steel, known for its excellent corrosion resistance, shock tolerance, and cost-effectiveness.

3. Prototyping to Production

Water Jet Cutting: Used for initial profiling to achieve precise outer shapes with minimal material waste. The plan was to keep using this tool for actual production thereafter due to the nice sanded edge it produced.

Testing & Iteration: Early versions underwent several rounds of real-world testing, particularly on how well nuts and bolts fit, leverage strength, and overall comfort in use. Each iteration answered specific performance questions.

4. Etching and Customization

Every Drais BMT is fiber laser-etched with permanent markings, including metric and imperial rulers. This process uses high heat and precision to ensure durability, markings that won't fade or wear off even with heavy use.

Custom branding, names, and logos can be added, making the BMT a popular option for business promotions and personalized gifts.

Challenges and Solutions

Balancing Strength with Cost
I needed a material that wouldn't compromise functionality but still kept production costs in check. 304 stainless steel hit that sweet spot.

Precision in a Small Package
Because of its compact size, even small tolerance variations could compromise fit and performance. Tight CNC waterjet standards were enforced to ensure secure hex openings and accurate fits for multiple tool functions.

Hydro-blasting stainless like a boss.

Without traction, even brilliance spins its wheels.

Scalable Customization

Creating a customizable product that could still be mass-produced required a reliable marking method. Fiber laser etching provided fast, high-detail personalization without sacrificing throughput.

Things I Had to Learn the Hard Way

Design for Real Use

A successful prototype isn't about flashy features, it's about surviving everyday wear, tear, and torque.

Simplify Without Sacrificing Utility

The challenge wasn't adding more tools, but refining the essentials that people actually need and use.

Precision Pays Off

Especially at small scales, accuracy isn't optional, it's the difference between a useful tool and a novelty.

Another One Off the Table

The Drais BMT wasn't just a clever bike tool, it was a full-blown journey from napkin sketch to foot-droppable prototype.

It's a great example of how thoughtful design meets precision manufacturing... and also a reminder that even the best ideas can sometimes go nowhere, no matter how solid the execution.

In the end, the project didn't take off, turns out it had no legs, but the process taught me a ton about turning flat sheet goods into something that could be custom-mass-produced with repeatable accuracy.

Sometimes, the real win isn't the product itself, it's the experience, the workflow, and the lessons you carry into the next build.

🏅 12.8 Practical Takeaways from the Shop

Prototypes won't lie to you: If something's wrong, they'll gleefully show you with a warped part or holes that don't line up. That's their job.

Blaming your CNC for a bad cut: Is like blaming the oven for burning your cookies when you set it to "volcano."

That gorgeous finish: Ruined by warping. That precise hole: Melted into a blob. Materials have personalities. Some are divas.

It might look great: Until you try to assemble it and realize it only fits together in Photoshop.

Prototypes don't lie. Reality's made of metal and dust.

❌ 12.9 Chapter 12 Quiz

Ready to ship or headed back to the shop? Let's put your confidence to the test.

1. What's the main goal of prototyping in manufacturing?
   A) Maximizing output volume
   B) Creating polished final products
   C) Avoiding the need for CAD
   D) Learning and refining designs

2. Which tool is most appropriate for prototyping?
   A) Waterjet cutter
   B) CNC mill
   C) Fiber laser
   D) Whatever works best

3. Pre-production vs. functional prototype?
   A) It's only visual.
   B) It's made using manual tools.
   C) It represents the near-final product.
   D) It avoids using CNC tools.

4. What's a common beginner mistake in prototyping?
    A) Treating the first prototype as the final product
    B) Skipping CAD entirely
    C) Ignoring aesthetics
    D) Using the wrong material on purpose

5. What follows the blueprint in prototyping?
    A) Client approval
    B) Rapid prototyping
    C) Material selection
    D) Final production

6. What's the biggest advantage of nesting in production?
    A) Makes products lighter.
    B) Enhances visual appeal.
    C) Reduces material waste and machine time.
    D) Speeds up manual design.

7. What's the solution for misaligned prototype holes?
    A) Increase material thickness.
    B) Recapture the CAD file with a photo.
    C) Use glue instead of mechanical fasteners.
    D) Recut the layout with adjusted spacing.

8. Which of these issues would waterjet cutting help avoid?
    A) Tool dulling
    B) Heat warping
    C) Alignment errors
    D) Surface scratches

9. Why are alignment tabs used in prototype redesigns?
    A) To simplify and improve assembly accuracy
    B) To make aesthetic symmetry
    C) To make the part lighter
    D) To avoid using glue

10. What's the solution for faint or unreadable engravings?
    A) Use stickers instead.
    B) Switch to screen printing.
    C) Adjust machine settings for deeper engraving.
    D) Paint over the marks.

11. Why should you test finishes during prototyping?
    A) To match competitor pricing
    B) To verify real-world durability and looks
    C) To save on machine time
    D) To avoid taxes on coatings

12. What does "repeatability" mean in custom production?
    A) Producing identical results every time
    B) Mass-producing thousands of identical units
    C) Always using the same material
    D) Never iterating on a design

13. What's the best tool for tight tolerances in metal?
    A) CO_2 laser
    B) CNC mill
    C) 3D printer
    D) Plasma cutter

14. What's an example of business-focused prototyping?
    A) Providing billable prototyping as R&D
    B) Using recycled scrap to build a prototype
    C) Skipping approvals to save time
    D) Hiding flaws from the client

15. Why did 304 made more sense than titanium for my tool?
    A) Titanium is too heavy.
    B) 304 is shinier.
    C) 304 balances strength with cost.
    D) Titanium is illegal to use.

16. Which CNC did I use to profile the Drais BMT?
    A) Fiber laser
    B) CNC router
    C) Drill press
    D) Waterjet cutter

17. Which process made the Drais BMT etched rulers?
    A) UV printing
    B) CO_2 laser
    C) Fiber laser
    D) Rotary engraving

18. What's key when optimizing a production workflow?
   A) Making parts in bulk
   B) Minimizing runtime, maximizing consistency
   C) Avoiding software
   D) Using as many machines as possible

19. What's a lesson from a warped prototype due to heat?
   A) Use heat-neutral tools like waterjets.
   B) Preheat the material before cutting.
   C) Always engrave first.
   D) Add glue to reinforce warped areas.

20. Why should you document workflows for recurring clients?
   A) So they pay faster
   B) To reuse designs across industries
   C) To reduce setup time and ensure consistency
   D) So employees don't quit

Small production run, waterjet cut, fiber etched.

✖ 12.10 Answer Key + Explanations

Here's where we reveal the magic, answer key with all the "how and why" explanations.

1D – *Prototyping is about answering questions and improving the design before committing to production.*
2D – *Conceptual prototypes are often visual, so fast, inexpensive tools are ideal.*
3C – *Pre-production prototypes are essentially the final version, used to validate before scaling up.*
4A – *Early versions are for learning, not perfection, mistakes are expected and necessary.*
5C – *Once a design is drafted digitally, material is chosen before any physical prototype is made.*
6C – *Nesting groups parts efficiently, minimizing offcuts and machine runtime across tools.*
7D – *Hole misalignment is best solved by tweaking your digital design and cutting again.*
8B – *Waterjet cutting avoids heat altogether, preserving delicate materials like acrylic or thin metals.*
9A – *Alignment tabs help ensure parts fit properly without guesswork or extra measuring.*
10C – *Changing engraving speed, power, or method can produce higher contrast and readability.*
11B – *Finishes aren't just for looks, they affect wear resistance, longevity, and usability.*
12A – *Repeatability ensures consistency, which is more valuable than volume in custom jobs.*
13B – *Mills offer the precision needed for complex internal features with tight tolerances.*
14A – *Early-stage work should be billed, it's value-added design, not free pre-sales work.*
15C – *304 stainless offers the best mix of performance and affordability for the use case.*
16D – *Waterjets cut with high precision and no heat, ideal for shaping metals efficiently.*
17C – *Fiber lasers are perfect for marking metal, offering durable, high-contrast etching.*
18B – *Efficiency in production means doing things faster, better, and the same every time.*
19A – *If a material warps under heat, switching to a non-thermal process prevents the issue.*
20C – *Documented workflows streamline repeat jobs, reducing errors and improving client satisfaction.*

Because your acrylic pottery embosser deserves a custom laser flex.

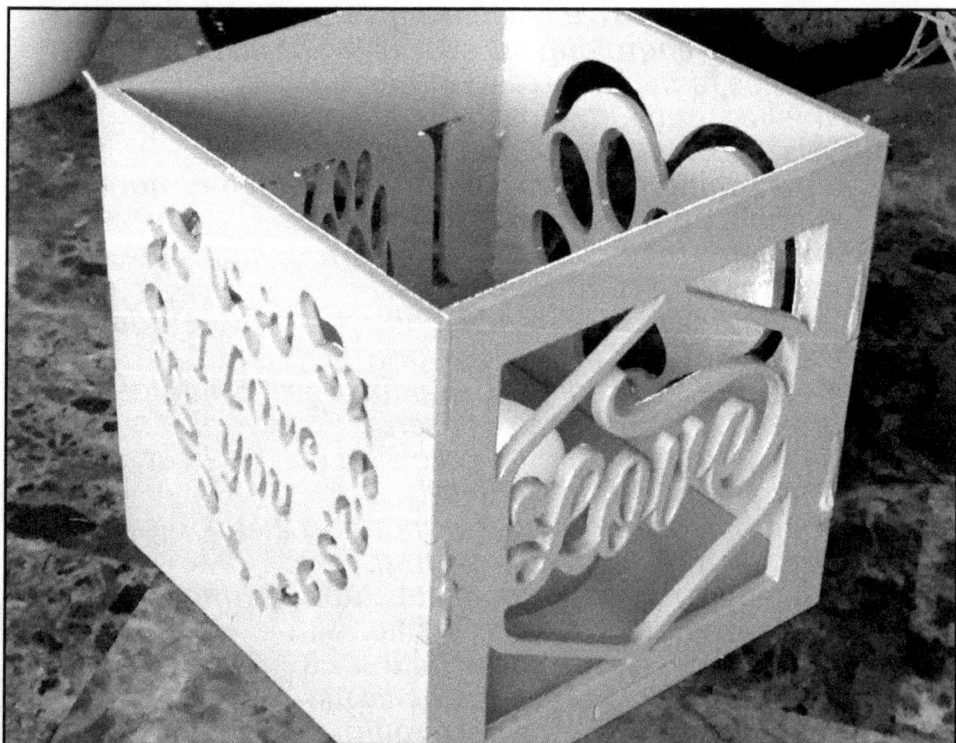

Mass produced custom tea lights was once a big business of mine.

Chapter 13
Finishing Techniques That Matter
Turning Ugly Ducklings Into Labeled Swans

No matter how accurate your CNC machine is, every part eventually needs a final touch, whether that's a tough coating, a crisp ID mark, or a bit of hands-on assembly. CNC does the heavy lifting, but finishing is what makes it shine.

A great finish puts your part front and center, impressing clients and setting you apart. A bad one? It gets quietly shelved, or worse, rejected outright.

This chapter covers those last, crucial steps, where precision meets presentation, and reminds you that all the effort chasing 0.0001 mm ($\approx$ 1/254,000") tolerances won't matter if you smudge the logo, or leave razor sharp edges.

Waterjet steel, gold paint, laser-engraved pine, sealed.

💬 *In custom manufacturing, finishing isn't just the final step, it's where the true craftsmanship and attention to detail shine through.*

13.1 Surface Finishing: Form Meets Function

Finishing isn't just about looks, it's about durability, feel, and performance. The right method depends on material, environment, and budget. Just remember, every finish adds time and cost, so plan accordingly in your quotes, or your budget might get a little too shiny for comfort.

Finish	Purpose	Common Materials
Sanding / Polishing	Smooth surface, prep for paint	Wood, plastic, metal
Anodizing	Corrosion resistance, color	Aluminum
Powder Coating	Tough, consistent, colored finish	Metals
Painting	Aesthetic, customizable	Wood, MDF, metal
Brushing / Graining	Directional finish	Stainless steel, aluminum
Clear Coating / Sealing	Protection and UV resistance	Wood, acrylic, composites

👎 *Overlooking finishing steps can ruin an otherwise great project.*

13.2 Marking: Identity, Instruction, and Branding

Marking is where parts get their identity, whether it's a logo, serial number, or barcode. From lasers to inkjet printers, each method has its perks.

Pick the right tool for the job to get precise, durable, and brand-worthy marks, because a part without a mark is like a movie without a title. Let's see which marking method suits your needs!

Marking	How It Works	Best On	Tool Used	Advantages	Disadvantages
Inkjet / UV Print	Adds full-color graphics or barcodes	Acrylic, plastic, coated metals	Specialty printer	Fast and flexible; can print full-color and complex graphics	Less durable; can fade or smudge over time; limited on some materials
Laser Engraving (CO₂)	Burns top surface or adds contrast	Wood, acrylic, leather	CO_2 laser	Clean, precise marks; great for non-metal materials	Limited depth; not suitable for metals
Laser Engraving (Fiber)	Ablates surface or deeply engraves metal	Stainless, aluminum, brass	Fiber laser	High precision on metal; deep engravings possible	Slower on non-metal materials; more expensive machine
Mechanical Engraving	Physically cuts into material	Wood, metal	Rotary engraving / router	Can create deep, durable markings; great for heavy-duty use	Can be slower; limited in fine details; may cause wear on tools
Stamping / Etching	Permanent but limited in detail	Metals	Press or acid etching	Very durable; permanent marking solution	Limited detail; only suitable for metals

13.3 Assembly Considerations: Where the Magic (and Mess) Happens

You've designed the perfect custom parts, great! But now comes the real test: assembly. This is where your project either comes together beautifully... or falls apart in your hands. Here's a look at the common challenges:

Alignment Issues: The design says the parts should fit, but in reality, they're playing hard to get. Gaps, misfits, and awkward angles are the price of tiny oversights.

Tolerance Stacking: One small error per part might seem harmless, but stack them up and suddenly your precise build looks more like abstract sculpture.

Hardware Conflicts: You thought you had the right bolts, but now a washer's too wide, or a nut won't fit. Delays, rework, and frustration ensue.

Human Factors Flaws: Sometimes the parts are fine, the problem is the person putting them together. Confusing instructions and fiddly components slow things down fast.

Assembly Tips to Keep Things from Falling Apart (Literally)

1. Design for Assembly (DFA)
Make life easier with tabs, slots, and alignment features. These guide the build and help eliminate guesswork.

2. Standardize Your Hardware
Stick to a few common fasteners wherever possible. Fewer sizes mean fewer chances to mess up, and fewer trips to the bin looking for "that one weird screw."

3. Dry Fit First
Before you commit to gluing, screwing, or welding, make sure everything fits together nicely. Materials like wood and acrylic can change shape when cut, so be careful.

4. Use Jigs

For precision or high-volume jobs, custom fixtures and jigs are your best friends. They boost accuracy and help you avoid that dreaded "oops" moment when something ends up just a hair off.

5. Document It

Don't leave the assembly process to guesswork, write it down. Step-by-step instructions and visual guides turn confusion into consistency. Your future self (and your team) will thank you.

By thinking ahead and using a few smart strategies, you'll move through assembly with fewer surprises, unless it's a surprise party with snacks. But that's another chapter.

Planning your builds with assembly in mind means fewer headaches, cleaner alignment, and less time digging for the right fastener. It's all about working smarter, not harder, because the only surprises you want during assembly are cake and confetti.

🖥 13.4 Don't Skip Finishing in Your Quote

1. Finishing adds time, materials, and complexity: So your pricing needs to reflect that. It's the first thing customers notice, so it should look great and be budgeted accordingly.

2. Powder coating costs vary by volume: Coaters charge per part or batch. Small runs are pricier per unit due to fixed setup time; larger batches are more cost-efficient as long as they are accurate.

3. Marking takes prep time: Variable data like serial numbers or logos require layout tweaks, testing, and alignment, especially with fiber lasers or engravers. That prep adds up.

4. Assembly can bottleneck fast: Misalign one part and the whole job slows down. Even simple builds need buffer time for fitting, fastening, and potential rework.

13.5 Cleaning and Quality Control: Because "Good Enough" Doesn't Cut It

Before you ship out your custom parts or start assembling them into the final masterpiece, it's time for a little TLC. Think of this step as the spa day for your parts, cleaning, polishing, and making sure everything is in tip-top shape. A solid Quality Control (QC) process isn't just a step; it's the difference between sending out a flawless product or an awkwardly crooked, dusty mess.

1. Deburring / Edge Cleanup

Burrs and sharp edges aren't just annoying, they're dangerous. Sharp edges can cause cuts, scratches, and even damage to your equipment. Plus, they make parts that should fit together nicely look like they've been through a wrestling match.

Tip: Get yourself a proper deburring tool or some sandpaper and smooth out those edges like you're a prepper.

Painted MDF then laser-engraved Lemarchand's box.

2. Cleaning (Solvent or Soap)

After machining, your parts are probably covered in oils, dust, and maybe a bit of mystery goo. You need to clean those off before any coating or finishing can happen. Otherwise, the paint will peel off faster than your last DIY project.

Tip: Use a lint-free cloth or compressed air to make sure you've wiped away every last spec of dirt before you go on to the fun stuff.

3. Visual Inspection

Your parts are looking good, but are they really looking good? Visual inspection helps catch those little burns, scratches, or weird warps that will make you question if a toddler assembled your project. Doubly so for welds.

Tip: Take a good, long look at both functional and aesthetic aspects, this is your last chance to catch any defects before moving on.

4. Fit Check

Parts should fit together like a glove. If you've got parts that need to be forced, it's like trying to dress a penguin for the desert, and that's no fun for anyone.

Tip: Test the parts before you get too committed. A good fit now means fewer headaches later on.

5. Protective Film Removal

That protective film is your part's safety blanket, but don't let it stay on too long! Removing it at the wrong time could leave scratches or greasy fingerprints that are harder to clean than your last takeout order.

Tip: Peel that film just before final assembly and use clean hands or gloves to avoid leaving your mark. I find thin cotton gloves to be the best when it comes to part handling.

13.6 Cleaning Tools to Keep on Hand

Stock up on these essential cleaning and QC tools, this is your bare-minimum kit for making parts look as good as they perform:

Lint-Free Cloths: Wipe down parts without leaving behind fuzz, fibers, or frustration.

Compressed Air: Blow dust out of sneaky corners and tight spots where cloths can't reach.

Solvents (or soap for non-critical parts): Use the right cleaner for the job. Acetone is a lifesaver for prepping metal surfaces before painting, just make sure it's compatible with your material. Windex is awesome for stainless.

Soft Brushes: Sometimes a gentle scrub is all it takes to clean up fine details or delicate finishes.

Nitrile Gloves: Protect finished parts from fingerprints, smudges, and mystery oils, especially during final inspection.

Investing time in cleaning and quality control means fewer do-overs, longer-lasting parts, and happier customers.

⬇ 13.7 Marking vs Engraving

When it comes to laser marking vs. engraving, it's a bit like choosing between a quick doodle and a carefully etched masterpiece. Both processes use lasers, but they work their magic in different ways, depending on what you need.

If you're after a quick identifier, like a logo or barcode, marking is your go-to. But, if you're aiming for something permanent and deep, like serial numbers or intricate designs, then engraving is the heavy hitter.

Just remember, lasers don't have a two-way z-axis feedback, so depth control isn't their strong suit.

📱 13.8 Success Story
Custom Plywood Boxes

Birch boxes getting ready for their flat-pack adventure.

CNCROi.com took on a large-scale project for a client needing hundreds of custom plywood boxes.

The challenge? Reduce shipping costs without compromising quality. So, we put on our thinking caps, and safety goggles, and got to work optimizing the design for flat-pack shipping using precise laser cutting, smart material use, and interlocking joints that made assembly quick, tool-free, and repeatable.

We also dialed in kerf allowances to ensure a snug fit, reducing the need for excessive amounts of glue.

The result? A set of functional, great-looking, and cost-efficient boxes that would make any shipping department proud, easy to transport, simple to assemble, and strong enough to protect what matters.

The Process

1. Design and Material Selection

We chose 1/8" (3 mm) birch plywood for its strength, clean look, and laser-friendly surface. The exterior got a light burnt finish for style, while the interior stayed plain. Tight tolerances helped prevent warping.

2. Efficiency Through Laser Cutting

Laser cutting delivered clean, precise cuts with minimal burning, perfect for thin plywood. We engraved directly on the surface, skipping masking to save time.

3. Optimization for Mass Production

Nesting three box patterns per sheet reduced waste and cut costs, supporting our eco-friendly approach.

4. Material Alternatives and Decision-Making

We considered MDF and stainless steel, but birch plywood struck the right balance of cost, strength, and appearance.

Things I Had to Learn the Hard Way

Precision Pays Off

Laser accuracy meant fewer errors, tight tolerances early on prevent major headaches later.

Optimize from the Start

Smart nesting saves both material and time. It's essential for large runs.

Material Selection Matters

Choose materials that do the job, enhance the look, and stay within budget.

Engrave it. Cut it. Lasered plywood's never looked this good.

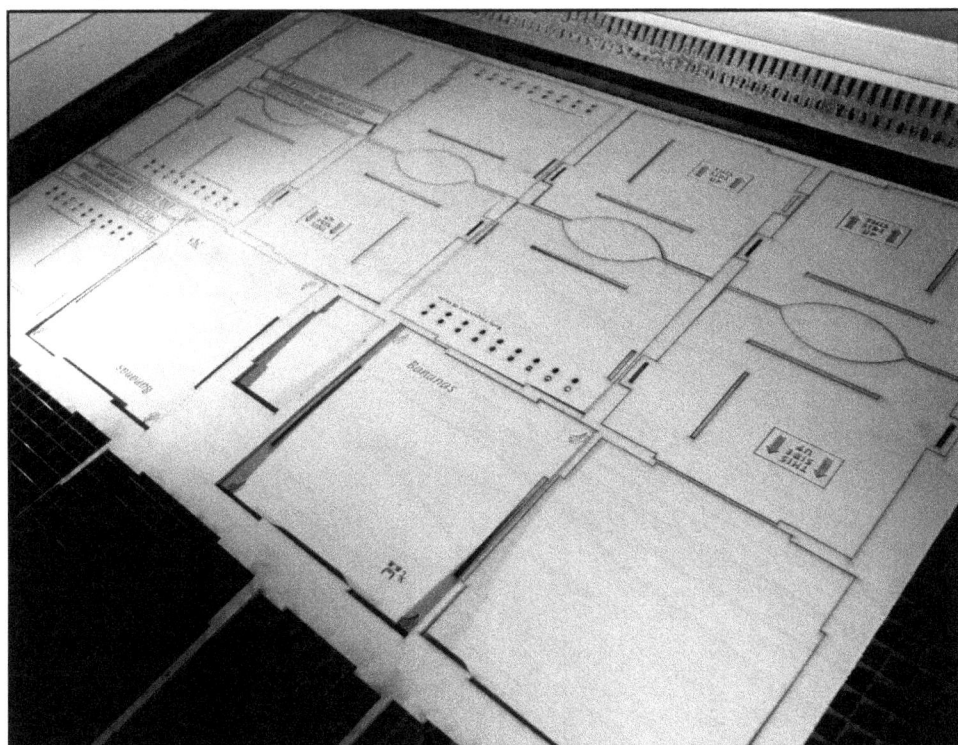

Lasers: Because plywood deserves both surgery and a signature.

Another One Off the Table

By using precise laser cutting and carefully selected materials, we delivered custom plywood boxes that cut shipping costs, simplified assembly, and gave the customer a top-tier, practical product, durable, easy to handle, and sharp enough to impress right out of the box. The flat-pack design not only saved space but also made logistics and storage more efficient.

🎖 13.9 Practical Takeaways from the Shop

Don't skimp on research: Choosing the wrong finish can sabotage an otherwise perfect part. It's like showing up to a black-tie dinner in sweatpants, technically functional, totally off the mark. Know your options: powder coating, anodizing, painting, plating, raw polish, each comes with its own strengths, limitations, and costs.

Finishing isn't free, and neither is your time: That "quick final step"? It's rarely quick. Prep, masking, drying, and curing all take time. Treat finishing as a real line item in your quote, or you'll end up eating the cost when deadlines get tight and margins get thin.

Skip cleaning and inspection, and you're asking for trouble: Surface contamination or hidden burrs can ruin adhesion, leave visible defects, or get your part rejected. A proper clean and thorough inspection before finishing saves time, money, and reputation.

Mockups are your friend: A small-scale or test finish can reveal problems before they become expensive disasters. It gives the client a preview and gives you time to fine-tune settings, confirm material compatibility, and lock in the process with confidence.

A flawless finish doesn't happen by accident. It's the result of planning, testing, and treating every step, from prep to polish, like it matters. Because it does.

❌ 13.10 Chapter 13 Quiz

This quiz is like a pop quiz, but with fewer cold sweats and more packing peanuts.

1. Why is surface finishing important in manufacturing?
   - A) To make parts look aesthetically pleasing
   - B) To improve the durability and usability of parts
   - C) To reduce manufacturing time
   - D) To increase the size of parts

2. Which material is anodized for corrosion and color?
   - A) Stainless steel
   - B) Aluminum
   - C) Wood
   - D) Acrylic

3. What's a common beginner mistake with finishing steps?
   - A) Spending too much on coatings
   - B) Adding too many layers of finish
   - C) Using the wrong materials for finishes
   - D) Overlooking finishing steps

4. What's the key advantage of powder coating?
   - A) It provides a tough, consistent, colored finish.
   - B) It is cheaper than painting.
   - C) It is ideal for wood and MDF.
   - D) It is suitable for non-metal materials.

5. What does CO_2 laser engraving work best on?
   - A) Stainless steel
   - B) Plastic
   - C) Wood
   - D) Aluminum

6. What's the best marking method for deep metal engravings?
   - A) Inkjet printing
   - B) Laser engraving (Fiber)
   - C) Stamping
   - D) Mechanical engraving

7. What's a common issue during manufacturing assembly?
   - A) Material shortages
   - B) Difficulty in finishing
   - C) Excessive labor costs
   - D) Alignment issues

8. What's the main benefit of designing for assembly (DFA)?
    A) Faster production of the parts
    B) More complex assembly
    C) Easier assembly and fewer errors
    D) Lower material costs

9. Why is dry fitting crucial during the assembly phase?
    A) To ensure that parts fit correctly before final assembly
    B) To test the strength of the material
    C) To optimize the coating process
    D) To adjust tolerances for better fitting

10. What's the role of jigs in the assembly process?
    A) To help clean parts before assembly
    B) To ensure accurate placement during assembly
    C) To cut material into smaller pieces
    D) To add additional features to parts

11. What's a key quality control step before assembly?
    A) Powder coating the parts
    B) Testing material strength
    C) Cleaning parts to remove oils and debris
    D) Adding final touches to aesthetics

12. Why is visual inspection crucial in quality control?
    A) It guarantees parts are within the required size.
    B) It ensures parts are clean.
    C) It checks for material cost efficiency.
    D) It identifies defective parts like scratches or burns.

13. What's the best tool for cleaning intricate parts?
    A) Compressed air
    B) Rotary tool
    C) Soft brush
    D) Solvent

14. Why should you wear gloves when handling parts?
    A) To avoid damaging the parts.
    B) To prevent contamination from oils and fingerprints.
    C) To keep parts from sticking to hands.
    D) To make handling parts easier.

15. What's a key consideration when quoting finishing?
    A) Finishing costs should not be included in quotes.
    B) The type of coating has no effect on pricing.
    C) Powder coating charges depend on volume.
    D) Assembly time is not a significant factor.

16. What's the best material for full-color laser marking?
    A) Wood
    B) Plastic
    C) Metal
    D) Acrylic

17. What's one of the benefits of mechanical engraving?
    A) It creates deep, durable markings.
    B) It is faster than laser engraving.
    C) It is ideal for non-metal materials.
    D) It does not require any tools.

18. What's the typical application for etching?
    A) Serial numbers on metals
    B) Decorative markings on plastic
    C) Logos on wood
    D) Barcodes on acrylic

19. Why should you avoid removing protective film too early?
    A) It could affect the durability of the part.
    B) It may cause scratches or fingerprints.
    C) It reduces the part's overall aesthetic value.
    D) It might weaken the part's structure.

20. What material was used for the custom plywood boxes?
    A) MDF
    B) Birch plywood
    C) Stainless steel
    D) Acrylic

Bonus Question: What's one reason to create a small-scale mockup before final finishing?
    A) It speeds up the cutting process.
    B) It helps test visual branding layouts.
    C) It reveals potential issues before full production.
    D) It makes packaging easier.

Finishing isn't only for wood, fiber etching needs cleaning too!

Keeping the firepit fed and happy with Birch scrap.

✖ 13.11 Answer Key + Explanations

Check your answers here and learn from the breakdown of each one!

1B – Surface finishing improves durability, feel, and function, not just looks.
2B – Anodizing adds corrosion resistance and color to aluminum. It's not used on wood or plastic.
3D – Skipping finishing steps often ruins a part's appearance and usability.
4A – Powder coating gives metals a tough, consistent, and often colorful finish.
5C – CO_2 lasers excel at engraving non-metals like wood, acrylic, and leather.
6B – Fiber lasers engrave metal deeply and precisely, making them ideal for stainless steel.
7D – Misalignment during assembly causes gaps, poor fit, and connection issues.
8C – Designing with assembly in mind reduces errors and speeds up the process.
9A – Dry fitting checks part alignment before final assembly, helping avoid issues.
10B – Jigs hold parts in place for proper alignment and fewer mistakes.
11C – Cleaning parts before assembly removes contaminants that could interfere with finishes or adhesion.
12D – Visual checks catch subtle defects like scratches or burns before the next step.
13C – Soft brushes clean delicate or detailed parts more gently than compressed air.
14B – Gloves prevent fingerprints and oils from contaminating the finish.
15C – Finishing services like powder coating are priced by volume, larger batches lower the per-part cost.
16D – Inkjet or UV printing works well for colorful designs on plastics or acrylic.
17A – Mechanical engraving creates deep, durable markings on metal parts.
18A – Etching (espcially lasers) are used for permanent IDs like serial numbers on metals.
19B – Removing protective film too early can cause scratches or smudges.
20B – Birch plywood was used for the custom boxes due to its strength, clean surface, and natural appearance.
BQC –Mockups reveal finishing or fit issues before full production, saving time, material, and headaches.

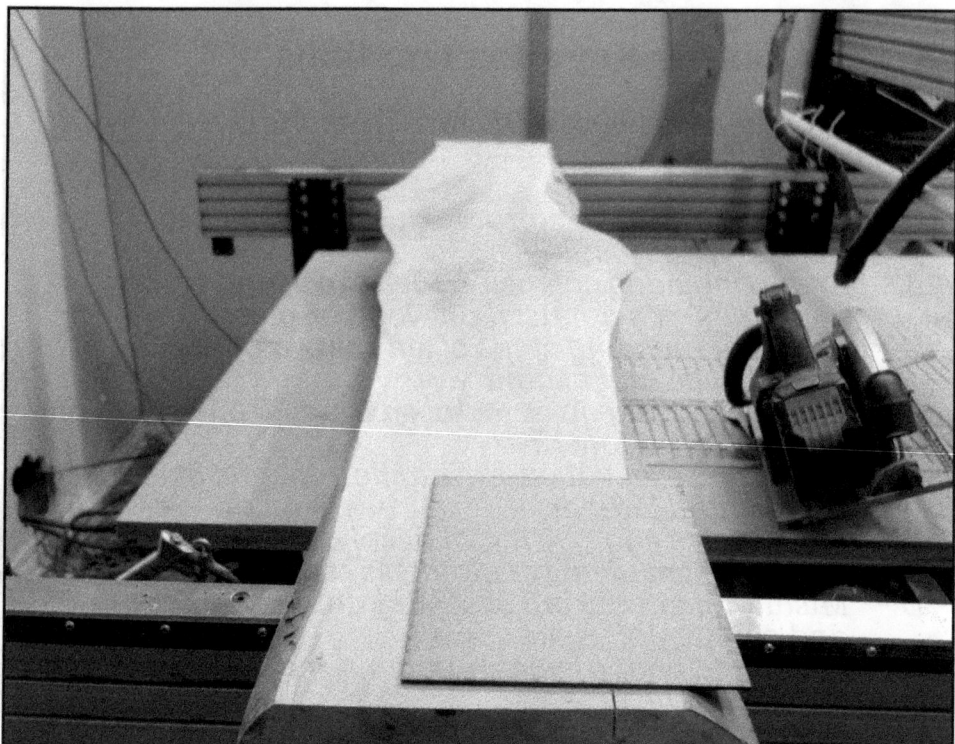
CNC laser template on sycamore live edge, classy meets crispy.

At a CNC shop, half the job isn't CNC, it's glue and grit.

Chapter 14
Packaging and Delivery Strategy
Ship Smart, Protect Craft, Impress Instantly

You've designed it, cut it, finished it, and checked it. Now comes one of the most underestimated stages in custom manufacturing: packaging and delivery.

This is where precision meets the chaos of shipping, handling, and sometimes impatient clients.

If it arrives scratched, bent, or confusing to assemble, it doesn't matter how perfect your tolerances were, it's a failure in the customer's eyes.

💬 *You can nail the design, the cut, and the finish, but if it shows up looking like it lost a bar fight, nobody cares.*

Sycamore awards, turning packing into a puzzle and a workout.

14.1 Packaging: Protection and Presentation

You have to assume your box will be dropped, tossed, stacked, rammed by a freight train, or even thown from a burning airplane in the middle of a hurricane, because it might be.

Packaging isn't just a protective buffer; it's an integral part of your product experience, not something to be considered as an afterthought.

Packaging Need	Solutions
Fragile parts (e.g., acrylic)	Bubble wrap, foam inserts, corner protection
Heavy items	Double-walled boxes, bracing, strapping
Moisture-sensitive parts	Desiccant packs, sealed bags, shrink wrap
Sharp edges / small parts	Edge protectors, zip bags, hardware pouches
Brushing / Graining	Directional finish
Branding	Custom labels, instructions, unboxing elements

14.2 Clear Documentation: Labels and Instructions

Even if you're delivering unassembled parts, clear documentation can save headaches for both you and your client.

When preparing kits or multi-part builds, label each component clearly to avoid assembly confusion. Include diagrams that show how parts fit together, with names or numbers that match the labels. For CNC or laser-cut parts, orientation guides or photos are especially helpful, mirrored or flipped pieces can easily cause mistakes.

Don't forget a packing list that includes quantities and material types. For extra support, consider adding QR codes that link to video walkthroughs or digital manuals.

👎 *Spending 40 hours perfecting the part... and about 4 seconds tossing it in a box like it's a clearance bin special.*

14.3 Delivery Methods: Speed and Costs

Always take pre-shipping photos, they're invaluable when it comes to insurance claims or resolving disputes. And regardless of your shipping method, overpack and overprotect. Never rely on a *"FRAGILE"* sticker to shield your package like a force field, most couriers treat that as more of a challenge than a warning.

Method	When to Use	Notes
Local Drop-Off	*Custom installs, large pieces*	*Plan for handling gear, site communication*
Courier (UPS/FedEx)	*Small-to-medium parts with tracking*	*Use insurance for high-value items*
Freight / Pallet	*Large or heavy parts, batch orders*	*Crate or strap required*
Flatpack Assembly	*Lightweight, large items (e.g., furniture kits)*	*Saves space, needs crystal-clear instructions*

14.4 Packaging for International Shipping

When crossing borders, packaging isn't just about protection, it's about compliance and communication.

Customs Documentation:

Skipping paperwork can lead to delays, fines, or even confiscation. Always double-check the requirements to avoid costly mistakes.

- Accurate commercial invoices listing contents, value, and use.

Packing awards: where "fragile" really means "good luck."

- Certificate of Origin (if required).
- Complete customs declaration forms (varies by country).
  International Regulations & Standards
- Wood Packaging: Must meet IPPC standards and bear approved stamps.
- Recyclability: Especially in the EU, packaging must often meet eco-standards.
- Language Requirements: Labels and instructions may need local-language translations.

Tamper-Proofing

With modern shipping, I rarely need specific tampering devices, but I always overpack items to prevent damage and reduce theft risk. Overpacking protects the contents and makes it harder for anything to be quietly removed.

When extra security is needed, adding seals, shrink-wrap, or holographic labels helps deter tampering and boosts customer trust by showing the package arrived as intended.

14.5 Environmental Considerations in Packaging

With sustainability becoming a top priority for some customers and corporations alike, eco-friendly packaging isn't just ethical, it's smart business.

Sustainable Materials
- Recycled paper and cardboard
- Biodegradable plastics (starch-based, compostable)
- Mushroom packaging: mycelium-based, shockproof
- Paper-based wraps instead of bubble wrap

Minimizing Packaging Waste
- Right-sizing boxes to avoid excess filler
- Reusable containers for high-value or repeat clients
- Zero-waste packaging (fully recyclable or compostable)

Eco-Friendly Printing & Labeling
- Soy-based inks (less toxic, easier to recycle)

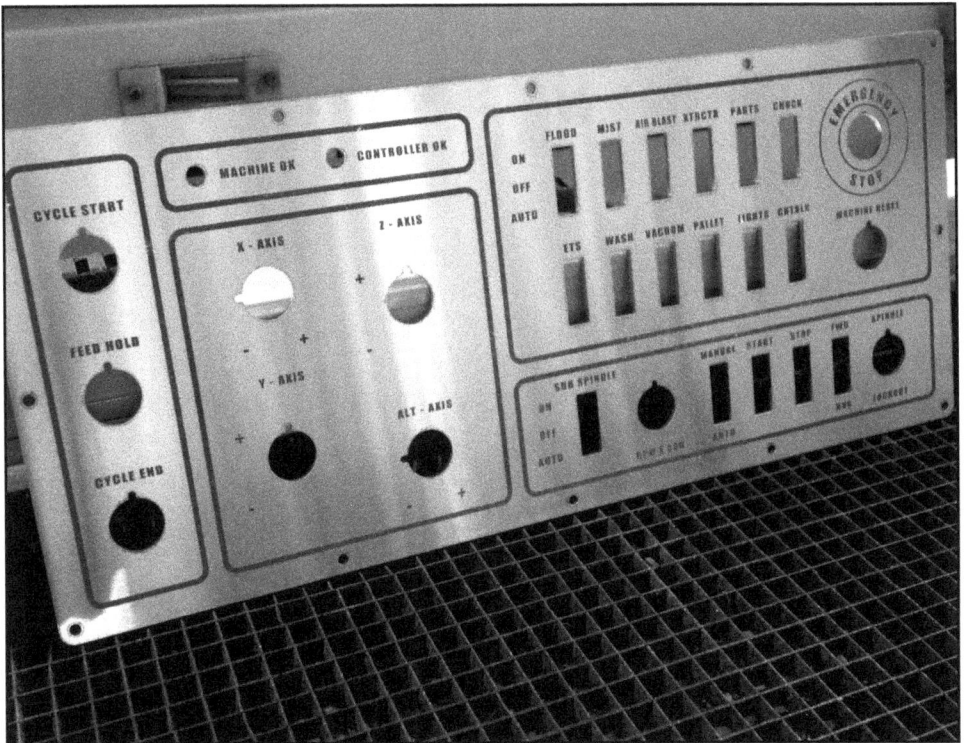

Indestructible things are never lost, just aggressively misplaced.

- Digital printing (low waste, efficient)
- Biodegradable or recycled labels, using water-based adhesives

Sustainability sells, clients notice when your values align with theirs.

14.6 Client Handoff: More Than a Transaction

This is your final touchpoint. Get it right, and you won't just ship a part, you'll earn a repeat customer.

The goal is for your customer to say, "Everything arrived in perfect condition, and the instructions made assembly a breeze." That's the kind of experience you want to deliver.

- Walk-throughs for installs or complex builds
- Follow up for feedback (and testimonials!)
- Provide handling/care tips or cleaning suggestions
- Include support and warranty info if relevant

🖥 14.7 Packing as Part of Your Brand

Underquoting packaging is a common mistake that quickly eats into profits. Packaging isn't just extra, it's part of the product experience. Good packaging:

- *Reduces damage and support calls: Protects parts, saving time and money on replacements.*
- *Saves you time: Fewer returns and remakes mean faster project turnover.*
- *Builds client trust: Clean, secure packaging shows professionalism and care.*
- *Costs money and time: Materials and packing labor aren't free, don't hide them in your pricing.*

Always itemize packaging and delivery in quotes and explain why they matter. It's not just a box, it's protection, presentation, and part of the product's value. Good packaging prevents damage, eases handling, and leaves a strong impression, reinforcing your professionalism.

☐ 14.8 Success Story
Custom Foam Inserts

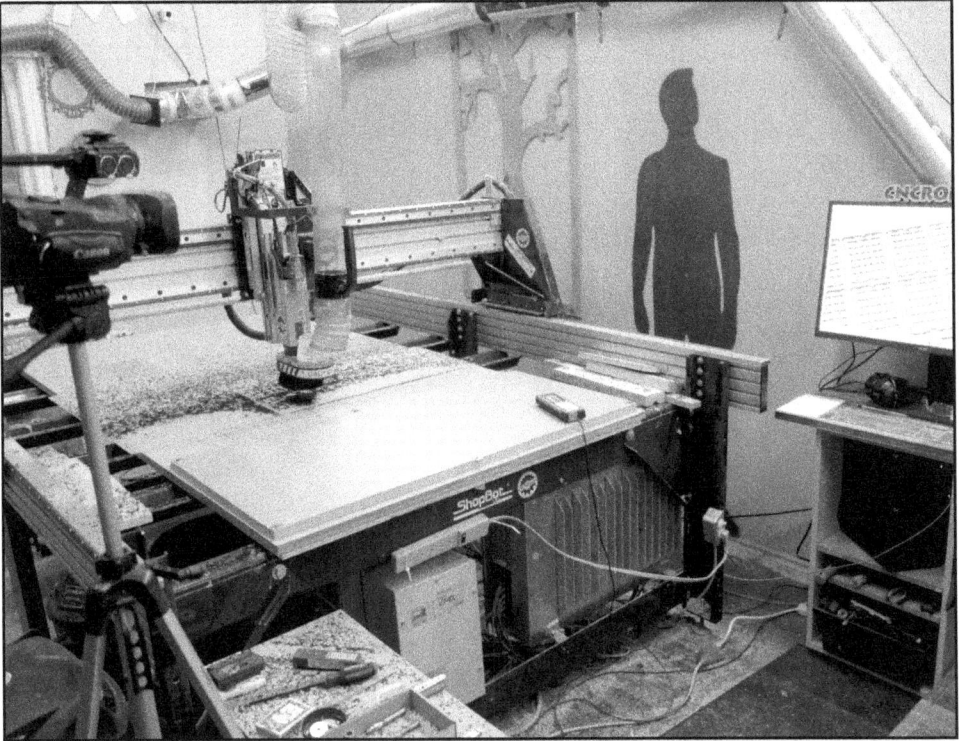

PRSalpha's fired up, miss this amazing CNC router!

At *CNCROi.com*, I had one goal: conquer foam with my ShopBot PRSAlpha CNC router and my usual mix of stubbornness and skill. Custom foam inserts? No problem. Precision? Always. But working with foam? It's like herding hyperactive dust bunnies that cling to everything except where you want them.

Process Breakdown

1. Initial Setup & Machine Specifications

My Shopbot PRSAlpha 4 ft x 8 ft was rigid, fast, and basically built to chew through anything... except foam, which isn't so much chewed as delicately whispered through. I've since done this job with my Thermwoods, which produced the same results. High precision meets low resistance. Sounds dreamy? Keep reading.

2. Challenges Encountered

Speed + Rigidity = Smooth Cuts (Usually):
This machine is overkill in the best way. With its rock-solid frame, it slices through foam like a hot knife through, well, foam. But don't let the softness fool you. Foam has a mind of its own, especially when static electricity shows up to the party.

One Tool to Rule Them All:
I kept things simple, just one tool for the entire job. Could I have used more? Yep. But I also enjoy not playing Russian roulette with flying bits. Predictability over speed wins on Day One.

Foam Snowstorm Prevention:
Our extraction system deserves a standing ovation. Without it, the shop would look like Frosty exploded. This thing keeps the static cling and debris from staging a full-scale rebellion.

Things I Had to Learn the Hard Way

Optimize Everything

Every project brings a fresh round of adjustments, cut speeds, feed rates, bit selection. It's less "set and forget" and more like fine-tuning a temperamental instrument. Except instead of music, it's foam dust flying in your face.

Get used to tweaking, testing, and dialing in settings every single time.

Foam is a Frenemy

It cuts like a dream but leaves behind a nightmare. Lightweight, static-prone, and clingy, foam loves sticking to everything except the vacuum hose.

Invest in a solid dust collection system, keep lint rollers on hand, and above all else, be patient. Cleaning up is part of the job, no matter how efficient your setup.

252

Efficiency Comes With Experience

The more you cut, the more you realize which tools you don't need. Streamlining your workflow, fewer bit changes, smarter toolpaths, optimized nesting, saves minutes that add up fast. Just know that chasing efficiency with foam means occasionally battling a static-charged blizzard of fluff. Welcome to the trade.

Another One Off the Table

At *CNCROi.com*, we turn foam frustration into functional brilliance. This project reminded me (yet again) that CNC is way more than just pushing a button, it's understanding materials, adapting processes, and occasionally yelling at a block of polyethylene like it's your nemesis.

Need custom foam inserts done right? You know who to call. (Hint: It's me.)

Cutting foam feels like cutting air, until you see the shipping bill.

🎖 14.9 Practical Takeaways from the Shop

Overprotect everything: Foam, bubble wrap, bracing, whatever it takes. That "FRAGILE" sticker? Couriers treat it like a challenge, not a cautionary warning.

Ship smart: Your business depends on it. Factor in size, value, fragility... and snap pre-shipping photos in case your package goes on vacation.

Pack responsibly: Use recycled, right-sized boxes and reusable materials. Customers notice, and so will your conscience.

Delivery isn't the end: It kicks off the next job. Make life easier with solid support, a smile, and a follow-up email that doesn't start with "oops."

❌ 14.10 Chapter 14 Quiz

Ready to wrap up your packing knowledge, or just wing it with confidence?

1. What's the primary goal of packaging in manufacturing?
    A) To make the product look aesthetically pleasing
    B) To protect the product during shipping
    C) To add weight to the product
    D) To create extra storage space

2. What's the best packaging for fragile acrylic parts?
    A) Bubble wrap, foam inserts, corner protection
    B) Heavy cardboard boxes
    C) Paper-based wraps
    D) Shrink wrap only

3. Which method is best for shipping large or heavy parts?
    A) Courier (UPS/FedEx)
    B) Freight/Pallet
    C) Local Drop-Off
    D) Flatpack Assembly

4. What ensures clarity in assembly kits?
- A) A clear product description
- B) A promotional flyer
- C) A receipt for the payment
- D) A packing list with quantities and material types

5. Why is overpacking important when shipping products?
- A) To make the product heavier for insurance purposes
- B) To reduce packaging costs
- C) To comply with international regulations
- D) To protect the product during shipping

6. What's the best way to ship moisture-sensitive parts?
- A) Edge protectors
- B) Bubble wrap
- C) Desiccant packs
- D) Shrink wrap

7. What's the best packaging for heavy items?
- A) Double-walled boxes with bracing and strapping
- B) Simple cardboard boxes
- C) Paper-based wraps
- D) Wood crates

8. What's the key factor in international shipping packaging?
- A) Ensuring the product is well-marketed
- B) Ensuring compliance with customs documentation
- C) Ensuring packaging is the cheapest possible option
- D) Ensuring the product is pre-assembled

9. What's important for international shipping packaging?
- A) Using wood packaging that meets IPPC standards
- B) Only using recycled materials
- C) Minimizing the use of labels
- D) Ignoring customs regulations

10. Which type of packaging is most eco-friendly?
- A) Recycled paper and cardboard
- B) Styrofoam peanuts
- C) Plastic bubble wrap
- D) Single-use plastics

11. What's the purpose of tamper-proofing in packaging?
    A) To make the packaging look more attractive
    B) To prevent tampering and build customer trust
    C) To lower the cost of shipping
    D) To comply with environmental standards

12. Which material is biodegradable and shock-absorbent?
    A) Styrofoam
    B) Bubble wrap
    C) Mushroom packaging
    D) Plastic packaging

13. Which language is required for international packaging?
    A) Labels and instructions may need local translations.
    B) Labels and instructions should only be in English.
    C) Only the product name needs translation.
    D) No language requirements needed.

14. Why should you itemize packaging and delivery in quotes?
    A) To increase the total cost of the job
    B) To reduce shipping costs
    C) To explain their importance to the client
    D) To avoid including packaging in the cost

15. What's the best way to ensure easy assembly for the client?
    A) Include a packing list.
    B) Send instructions only upon request.
    C) Ship the product unassembled.
    D) Include assembly diagrams with part labels.

16. Why should you use eco-friendly printing and labeling?
    A) To reduce printing costs
    B) To attract environmentally conscious clients
    C) To make the product heavier
    D) To comply with legal requirements

17. What's the challenge in CNC routing foam?
    A) Foam debris and static cling
    B) Excessive noise during operation
    C) Lack of precision
    D) Difficulty with cutting through harder materials

18. How does *CNCROi.com* prevent foam debris buildup?
    A) By using a high-speed router
    B) By cutting foam manually
    C) By implementing a robust extraction system
    D) By reducing the rigidity of the router

19. What does the foam insert case study reveal about CNC?
    A) It is not ideal for foam material.
    B) It requires multiple tools to optimize the process.
    C) It works best with metals and hard materials.
    D) It can cut foam at high speeds with ease.

20. What's the goal of client hand-off in manufacturing?
    A) To ensure that the product arrives on time
    B) To impress the client with the packaging
    C) To ensure easy client assembly and use
    D) To minimize the cost of shipping

Foam on my Thermwood CS43 stick to everything... except the job!

Customs must open my packages and wonder, 'What's this?'

✖ 14.11 Answer Key + Explanations

See if your shipping game delivered, here's the answer key with quick explanations!

1B – Packaging protects the product from damage during shipping.
2A – Fragile items like acrylic need cushioning to absorb impact.
3B – Heavy or large items ship best via freight with crates or strapping.
4D – Clear documentation ensures smooth assembly, especially for kits.
5D – Overpacking protects against courier mishandling during transit.
6C – Moisture-sensitive parts need desiccants to prevent damage.
7A – Heavy items require double-walled boxes and bracing for support.
8B – Proper customs paperwork prevents delays and fines in international shipping.
9A – Wood packaging must meet pest-control standards for global compliance.
10A – Recycled materials are eco-friendly and widely accepted.
11B – Tamper-proof packaging increases safety and customer confidence.
12C – Mushroom-based packaging is biodegradable and shock-absorbent.
13A – Some countries require local language labels for imported products.
14C – Itemizing packaging and delivery shows their value to clients.
15D – Clear instructions and diagrams reduce assembly errors.
16B – Eco-friendly printing supports sustainability and appeals to green buyers.
17A – Foam creates static, clinging to tools and leaving debris behind.
18C – A strong extraction system keeps foam mess under control.
19D – CNC router rigidity allows fast, clean foam cutting.
20C – A smooth client hand-off ensures easy assembly and reliable results.

Don't just "make" rubber stamps and ship them out...

Test them out first and show proof to the client it works!

Chapter 15
Future-Proofing CNC Workflow
Workflows That Evolve, Not Expire

Technology moves faster than a CNC spindle with the feed cranked too high. But custom manufacturing? It evolves like a chess grandmaster, strategically, patiently, and with zero tolerance for shiny object syndrome.

Staying ahead isn't about grabbing every flashy gadget; it's about knowing when to upgrade, what to sideline, and how to fold new tech into your workflow without flipping your entire shop upside-down.

Let's take a peek into the not-so-distant future, where AI helps you optimize, cobots don't steal your job (they help you lift things without herniating), and magnetic levitation is part of the everyday.

I've come a long way since my Model-T design in 2011.

💬 *Future-proofing your workflow isn't about chasing the latest gadget, it's about knowing when to experiment and how to adapt without losing your edge.*

15.1 The Rise of Cobots & Automation

As mentioned in Chapter 11, collaborative robots, "cobots", are no longer science fiction; they're now a practical part of modern workshops. Whether in small-scale fabrication shops or high-tech manufacturing lines, cobots have quietly become indispensable.

Unlike the old industrial robots locked behind safety cages and programmed for repetitive, isolated tasks, cobots are designed to work alongside humans, not just replace them.

They're built with sensors, force feedback, and safety protocols that make close human interaction not only possible but efficient and safe.

They enhance workflows by combining human skill with automated precision. While robots excel at repetitive, high-accuracy motions, humans bring adaptability, decision-making, and creative problem-solving to the table.

Task	How Cobots Help
CNC router / laser loading	*Consistent part placement, fatigue-free operation*
Waterjet head swapping	*Precision tool changes with minimal error*
Part bin sorting (post-cut)	*Reduction of sorting/handling mistakes*
Welding (GMAW/ GTAW/FCAW/ MCAW)	*Continuous bead quality, 24/7 productivity*

👎 *Chasing the Latest Trends Without a Clear Strategy*

15.2 AI in Custom Manufacturing

Sure, AI can churn out endless cat videos and suspiciously catchy ad jingles, but in the workshop, it's got bigger bolts to tighten. Today's AI isn't here to replace you with a robot that steals your job. Like its brawny cousin the cobot, AI is here to help, spot patterns, optimize cuts, reduce waste, and suggest improvements you didn't know you needed (but now can't unsee). It's less "Skynet" and more "Shop Assistant That Never Sleeps."

Think of AI as your digital co-pilot, one that's surprisingly good at math, doesn't need breaks, and won't complain about the smell of burnt MDF.

Real-World AI Applications in the Workshop

Nesting Optimization: AI-based software can reduce material waste by dynamically arranging parts for max yield, especially useful across mixed materials.

Predictive Maintenance: AI + sensors = alerts before something breaks. Downtime killer!

Quote Generation: Machine learning models analyze previous jobs to improve quoting accuracy and speed.

Design Assistance: AI-driven topology optimization can suggest lighter, structurally sound designs, ideal for performance-critical components.

15.3 Next-Gen Tech: What's on the Horizon?

Welcome to the edge of what's possible. New technologies are hitting workshops, set to transform how we build, weld, route, cut, and design.

Smarter automation, AI optimization, advanced materials, and cleaner, faster cutting are making it all about working smarter. What once took hours will soon take seconds, and the impossible will become routine. In many ways, the bricks have been layed, just the house is incomplete.

Magnetic Levitation Rails: Frictionless Motion for speed

Borrowed from futuristic transportation, maglev rails use opposing magnets to eliminate friction. In CNC machines, this translates to smoother, faster, and more precise motion. Already making their way into high-end 5-axis CNC systems, maglev technology could soon redefine ultra-precision fabrication in fields like aerospace, sculpture, and advanced component manufacturing.

Faster Speeds: Boosting throughput by removing drag.

Lower Maintenance: Less wear = less downtime.

Incredible Precision: No mechanical drag = more accurate motion.

Hybrid Machines: Multiple capabilities in One machine

Why switch between machines when one hybrid unit can do it all? Ideal for prototyping and small-batch runs, these all-in-one systems are advancing rapidly, and they're becoming more affordable than you'd expect.

Laser + CNC Router: Cut and engrave with precision and depth.

3D Printer + 5-Axis Mill: Additive + subtractive = complex, high-precision parts.

Benefits:
- Reduced floor space
- Seamless transition between fabrication types
- Shorter production cycles

The more processes we can shift into a "lights-out" environment, the greater the consistency, precision, and predictability we can achieve, without anyone needing to babysit the machine at 2 a.m. *Yes, Simon was ahead of the times, he just forgot his shop wasn't running off AI just yet, living in the future but using yesterday's technology.*

Arc time beats screen time, every time.

Fiber Laser Welding: Speed meets strength

Where traditional GMAW and GTAW welding fall short, like with thin metals, precise seams, or tight heat control, fiber laser welding shines. Once reserved for aerospace and medical applications, it's now within reach for custom fabricators aiming for top-tier results.

Advantages:
- Reduced Distortion: Less warping on heat-sensitive materials.
- Faster Welding: Higher power density = quicker joins.
- Cleaner Seams: Minimal spatter = minimal cleanup.

Smart Materials: Reactive Materials

From spacecraft to smartwatches, responsive materials are finally making their way into the workshop. Think adaptive furniture joints, self-tightening fixtures, and wear-resistantprototypes, once futuristic, now increasingly standard. Welcome to the era of responsive design.

Key Examples
- Shape Memory Alloys (SMAs): Metals that "remember" and return to a set shape when heated.
- Self-Healing Polymers: Plastics that repair surface damage, ideal for high-wear components.

15.4 Stay Curious: Experiment to Stay Ahead

Future-proofing isn't about chasing the latest gadget, it's about recognizing what could work better and knowing the right moment to test it.

That's why experimentation should be a core part of your workflow. The real advantage lies in knowing when to say "yes", and to what.

- **Innovation = Efficiency:** Even small improvements from fiber welding or nesting AI can add up over thousands of parts.
- **Exploration = Differentiation:** Trying smart materials or hybrid setups can unlock entirely new product lines.
- **Risk = Opportunity (when controlled):** Try on a small batch, evaluate, then scale what works.

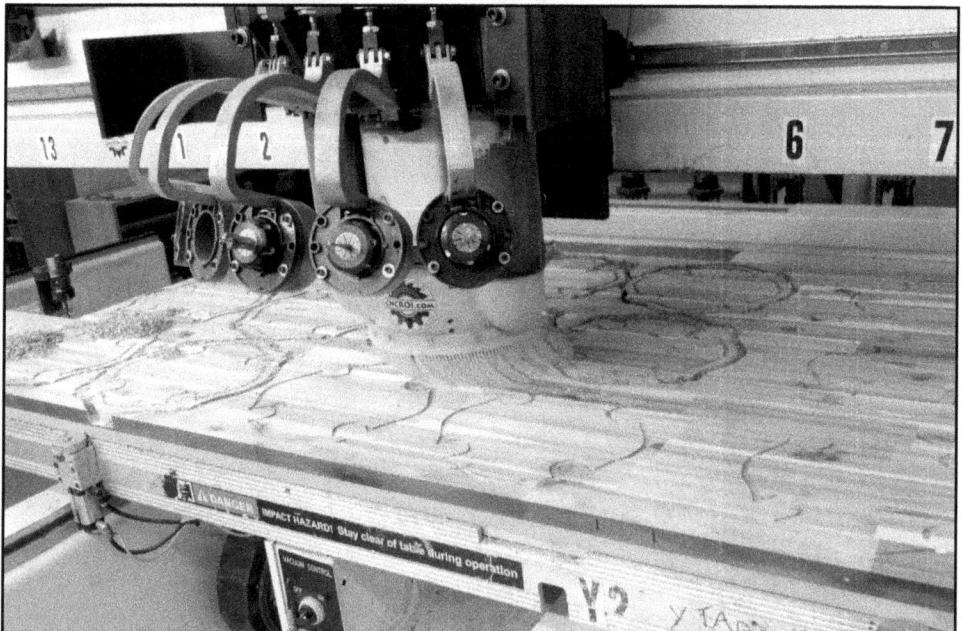

Big projects grow your mind, and your business.

15.5 Sustainability & Circular Design

Sustainability isn't just some trendy buzzword slapped onto packaging, it's a business no-brainer. The mission? Build smarter, cleaner, and longer-lasting. Think less landfill, more legacy.

Circular Design: Cradle-to-Cradle, Not Cradle-to-Landfill

Forget the old "build, use, toss" mentality. Circular design says: build it to last, fix it when it breaks, and reuse what you can until even the recycling bin gets jealous.

Modular Design: Snap-in, snap-out parts. Like LEGO, but with fewer foot injuries.

Disassembly-Friendly: Screws over glue, because glue is forever, and not in the good way.

Built to Last: Better materials = fewer replacements = fewer annoyed customers.

🖥 15.6 Why Does Sustainability = Smart Business?

Going green isn't just about hugging trees (though we're not against it). It's about running a smarter, cleaner shop with loyal customers, a stronger brand image, and lower operating costs.

Save Money: Reuse scrap, reduce power consumption, and avoid "green" surcharges. Efficiency pays off in material, energy, and disposal savings.

Future-Proof Your Shop: Environmental regulations are tightening. Making changes now avoids fines, delays, and expensive last-minute upgrades.

Win Customer Loyalty: Buyers increasingly prefer eco-conscious brands. Show them you care about the planet, and they'll reward you with trust, and repeat business.

15.7 Off-Planet Manufacturing: CNC Goes Lunar

While CNC machines have been reshaping manufacturing here on Earth for decades, projects like ICON's lunar construction initiative prove that CNC technology isn't just grounded, it's ready for lift-off.

In collaboration with NASA, ICON is developing massive additive CNC systems designed to 3D print entire habitats on the Moon, using regolith, the Moon's natural supply of loose rock and dust (basically extraterrestrial concrete mix).

And no, this isn't the plot of a sci-fi movie. This is fully automated, remote-controlled, extraterrestrial fabrication at its finest. ICON's robotic systems are engineered to survive harsh lunar conditions: extreme temperatures, radiation, micrometeorites, and absolutely zero on-site tech support.

These autonomous builders are meant to be delivered, powered up, and left to work on their own, building layer by layer without a human even setting foot on the site.

The implications for manufacturing back on Earth are just as groundbreaking. If we can 3D print buildings in a place where sunlight and shade can swing hundreds of degrees Celsius (or Fahrenheit) in hours, we can certainly build faster and more affordably in remote, dangerous, or disaster-stricken areas here.

It points toward a future where CNC manufacturing isn't just customizable, it's scalable, self-reliant, and deployable across planets.

ICON's lunar project captures the ultimate dream of CNC: designing and fabricating solutions anywhere, whether it's a crater on the Moon or the middle of nowhere on Earth, with no coffee breaks required.

For more information, visit:
https://www.iconbuild.com/lunar-construction

🏅 15.8 Practical Takeaways from the Shop

Don't chase tech blindly: Grabbing every shiny upgrade without a plan turns your workflow into chaos. Adopt with intent, not impulse.

Test before you commit: Small trials keep your shop efficient and your budget happy. Like sushi, try a piece before you order the boat.

Keep people in the loop: Automation's great... until it misplaces your best part. Smart humans are still your best backup.

Think sustainably early: Skipping sustainability now is future regret. Circular design saves money, earns trust, and avoids headaches later.

❌ 15.9 Chapter 15 Quiz

Think you're a brainiac? Take this quiz and prove it (or not)!

1. What's a collaborative robot (cobot)?
     A) A robot that works alongside humans.
     B) A robot that works independently.
     C) A robot that replaces human workers.
     D) A robot designed for industrial purposes only.

2. How do cobots enhance CNC part loading?
     A) They increase the load time.
     B) They prevent material damage.
     C) They reduce machine speed.
     D) They ensure consistent part placement.

3. Which AI reduces material waste in manufacturing?
     A) Predictive maintenance
     B) Nesting optimization
     C) Design assistance
     D) Quote generation

4. What's the main benefit of AI predictive maintenance?
   A) It increases machine speed.
   B) It reduces the need for operators.
   C) It alerts users before machine failure.
   D) It reduces material waste.

5. What does maglev rail technology offer CNC machines?
   A) Increased material waste
   B) Increased friction for better grip
   C) Enhanced operator control
   D) Frictionless motion for precision

6. What's one key benefit of hybrid machines?
   A) They save space and boost production cycles.
   B) They are easier to use than traditional machines.
   C) They only perform one task.
   D) They are slower than single-purpose machines.

7. Which welding tech is fast, precise, and low distortion?
   A) GMAW welding
   B) Fiber laser welding
   C) SMAW welding
   D) Arc welding

8. Which smart material 'remembers' its shape when heated?
   A) Shape Memory Alloys (SMAs)
   B) Self-Healing Polymers
   C) Thermoset Plastics
   D) Carbon Fiber

9. Why use self-healing polymers in custom manufacturing?
   A) To make materials stronger
   B) To repair surface damage automatically
   C) To improve heat resistance
   D) To increase flexibility

10. What does 'circular design' mean in manufacturing?
   A) One-time use of materials
   B) Rapid prototyping
   C) Maximizing energy consumption
   D) Building for reuse and regeneration

11. How does circular design support sustainability?
     A) By reducing material reuse
     B) By making products less durable
     C) By focusing on modularity and recyclability
     D) By ignoring product lifecycle

12. Why is sustainability smart for custom manufacturing?
     A) It reduces the need for skilled workers.
     B) It leads to cost savings and customer loyalty.
     C) It increases production time.
     D) It eliminates the need for new materials.

13. What's the benefit of fiber laser welding in manufacturing?
     A) More spatter and cleanup
     B) More power consumption
     C) Faster welding and reduced distortion
     D) Longer production times

14. What's the purpose of AI-assisted design assistance?
     A) To reduce machine downtime
     B) To simplify the production process
     C) To increase product complexity
     D) To create performance-optimized designs

15. What's the benefit of hybrid machines for small-batch runs?
     A) They offer flexibility and reduce machine needs.
     B) They increase production time.
     C) They are cheaper than single-purpose machines.
     D) They require fewer skilled operators.

16. How does AI-driven topology optimize manufacturing?
     A) It generates more complex designs.
     B) It suggests lighter, structurally sound designs.
     C) It reduces energy consumption.
     D) It prevents material waste.

17. What's a key feature of modular design in sustainability?
     A) Complex assembly processes
     B) Difficulty in disassembly
     C) Easy-to-replace parts for longer life
     D) Single-use components

18. How do cobots assist in welding (GMAW/GTAW/FCAW)?
    A) They reduce the number of welds needed.
    B) They ensure consistent beads and 24/7 productivity.
    C) They make the process slower.
    D) They replace human welders entirely.

19. What defines AI-based quote generation?
    A) It increases the time required to generate quotes.
    B) It provides only rough estimates.
    C) It eliminates the need for human estimators.
    D) It helps with quoting accuracy based on past jobs.

20. What's the major advantage of AI in manufacturing?
    A) It replaces all manual labor.
    B) It creates more manual tasks.
    C) It boosts decision-making and efficiency.
    D) It reduces product quality.

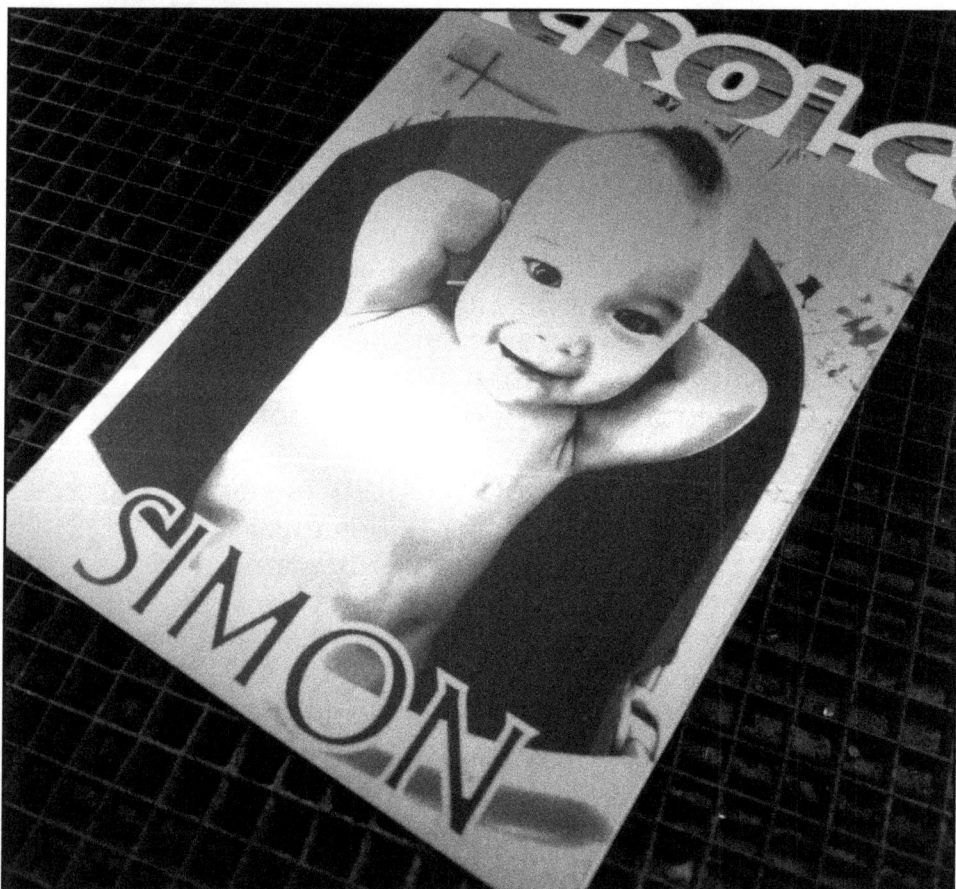

*Simon flexing on 316SS #4 finish, **PURE POWER!***

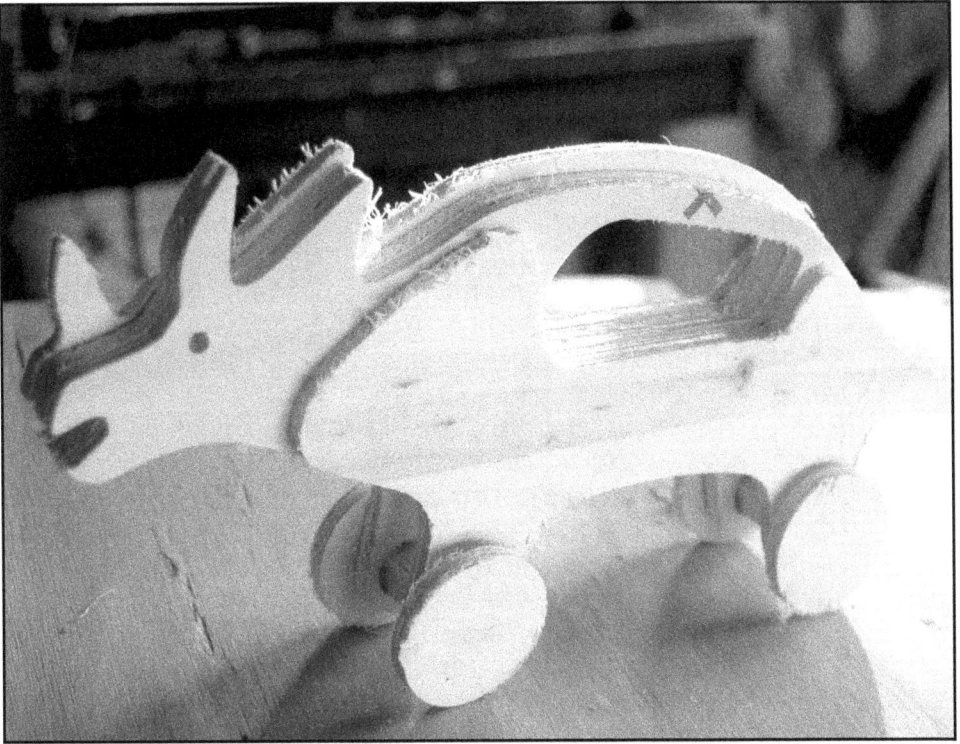

This took me over a week to design...

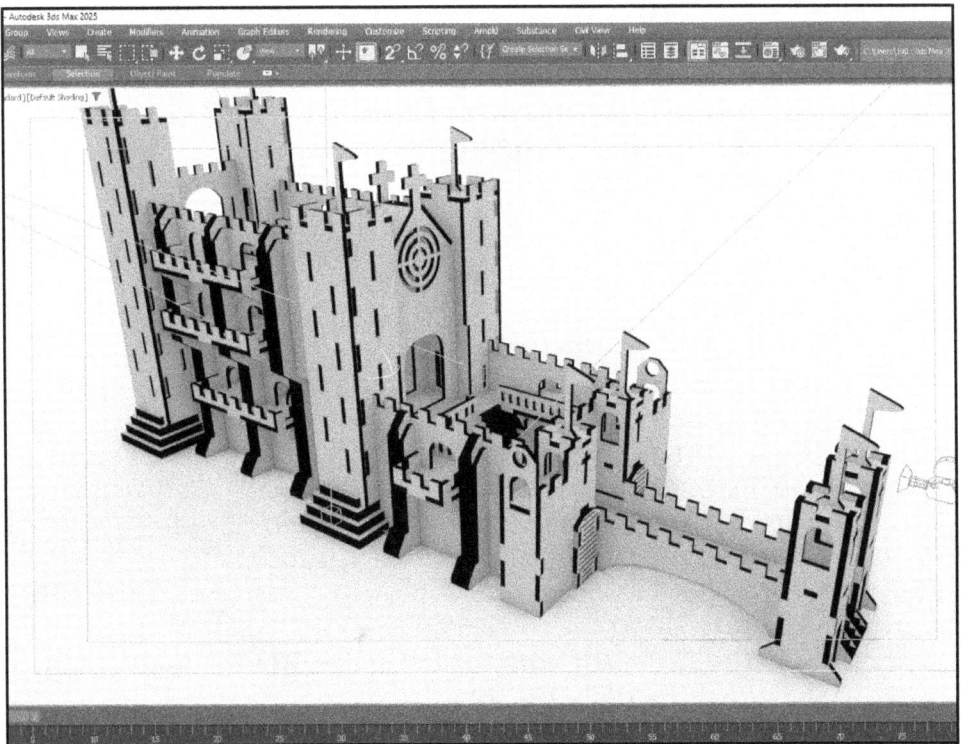

This took me two... a few years later.

✖ 15.10 Answer Key + Explanations

Here's the answer key, complete with all the explanations so you can really get it!

1A – Cobots assist humans, improving productivity without replacing workers.
2D – Cobots maintain consistent part placement, boosting efficiency and reducing fatigue.
3B – AI-driven nesting minimizes waste and improves material yield.
4C – Predictive maintenance uses AI and sensors to detect issues early, cutting downtime.
5D – Magnetic levitation eliminates friction for smoother, faster, and more precise CNC movements.
6A – Hybrid machines combine multiple capabilities, saving space and streamlining workflow.
7B – Fiber laser welding offers faster speeds, better control, and less distortion than traditional methods.
8A – SMAs return to a set shape when heated, enabling responsive applications.
9B – Self-healing polymers repair surface damage automatically, extending product life.
10D – Circular design aims for products that are reusable, recyclable, or regenerable.
11C – Circular design encourages products that can be disassembled and reused, reducing waste.
12B – Sustainable practices cut costs and build loyalty among eco-conscious buyers.
13C – Fiber laser welding is ideal for fast, low-distortion work on heat-sensitive materials.
14D – AI-driven design assistance generates structurally sound, efficient designs.
15A – Hybrid machines are perfect for small-batch runs, saving space and increasing flexibility.
16B – AI-driven topology optimization creates lightweight, strong structures.
17C – Modular design allows easy part replacement, extending product life and sustainability.
18B – Cobots ensure consistent welding quality and operate continuously to boost productivity.
19D – AI-driven quoting improves the speed and accuracy of estimates based on past jobs.
20C – AI improves decision-making and streamlines workflows in custom manufacturing.

Lots of love in this picture (laser cut 1/8" or 3 mm acrylic hearts).

Custom BBQ grill plasma cut using 12 ga 304SS 2B finish.

Final Words
The Build Never Really Ends
Stay Curious, Keep Creating, and Iterate Boldly

Congrats on crossing the finish line! The toughest part of writing a book? Deciding when to stop, especially when you've got a never-ending list of things you could add. I could easily keep going, but hey, I'll save that for the next Q&A focused book in this series. Whether you're just getting started or already halfway to becoming a CNC wizard, here's the real takeaway:

Success in custom manufacturing isn't about having the flashiest machine, the cheapest prices, or a shop that looks like the inside of a spaceship. It's about mastering your processes, knowing your materials inside and out, and consistently turning your ideas into more than just an expensive pile of scrap.

Sure, bigger investments often mean better results, but it's the smart investments that make you look like a genius (or at least save you from explaining a very expensive learning experience to your accountant).

And let's be honest: the upfront cost of a CNC machine is nothing compared to the time, creativity, and problem-solving you'll pour into it. CNCs are not magic boxes that print money while you zone out admiring your genius. They are curiosity-powered beasts fueled by innovation, experimentation, design, and a whole lot of trial and error. Your clients won't just expect you to cut parts; they'll expect you to anticipate their needs, spot potential issues, and deliver solutions they couldn't even articulate.

Over the years, CNC technology has evolved at a breakneck pace. What once required massive rooms full of specialized operators can now fit on a benchtop and be managed from a smartphone. We've moved from manual tool changes and G-code typed line by line, to fully integrated, touchscreen-driven systems that practically set themselves up.

CAM software that once required days of manual calculations now optimizes toolpaths in minutes, and that's just scratching the surface.

Today, we're standing on the edge of another leap: artificial intelligence, cobots, modular machines, and adaptive real-time manufacturing. CNC systems are already learning to optimize themselves, adjust on the fly, predict maintenance needs before a failure happens, and even suggest better designs before you hit "run." **The future isn't just automation, it's autonomous decision-making.**

And while your CNC might not need motivational posters in the shop, the machines your kids will work alongside might actually offer you a prep talk on Monday mornings, before asking if they can optimize the next batch run.

The world my son will grow up in will be very different from the one I started my journey in. Hopefully, it will be a world where the incredible power of CNC, combined with AI, allows creativity to soar even higher. A world where machines don't replace people but empower them to do more, think bigger, and solve problems faster than ever before. If we steer it right, it will be a future where custom manufacturing isn't just about making things, it's about imagining better ways to live, work, and build together.

I hope this book has given you not just fresh insights into the world of CNCs, but a real sense of the excitement and opportunity that lies ahead. These amazing machines are slowly taking over our lives behind the scenes in shops around the world, for the better.

And trust me: the journey's just getting started. Buckle up. CNCs with AI are about to learn at an exponential rate, and they're bringing you (and the next generation) along for the ride.

I can only hope that I am fortunate enough to experience some of these remarkable breakthroughs firsthand.

Q&A
Your CNC Questions Answered
Real Solutions for Real-World Challenges

Q1: What's the difference between GMAW and FCAW welding?

GMAW (Gas Metal Arc Welding), also known as MIG welding, though, let's be honest, it probably should be called MAG since the gas is active, not inert most of the time, uses a solid wire electrode and needs an external shielding gas like argon or CO_2. This makes it perfect for clean environments and thin materials.

Now, FCAW (Flux-Cored Arc Welding) is the rebel of the welding world. It uses a flux-filled tubular wire and, in some versions, doesn't require shielding gas, making it the go-to choice for outdoor or fieldwork and thicker steel.

So, here's the deal: GMAW requires shielding gas, uses a solid wire, and isn't ideal for outdoor use because wind can blow away your precious gas. It delivers clean welds and good penetration, if you know what you're doing.

FCAW, on the other hand, can run with or without shielding gas, uses a flux-cored wire (hence the name), and is much more outdoorsy. It's better for thicker materials but comes with the price of more cleanup due to spatter and slag.

And just to clear up a common question, no, you can't use shielding gas with self-shielded FCAW (Flux-Cored Arc Welding) or skip the gas with gas-shielded FCAW. Trust me, trying to mix those up is like trying to make a sandwich without bread. The flux mix is optimized for each type.

Self-shielded flux-cored wire works by reacting with the surrounding air to protect the weld. Add gas, and it messes up the whole recipe.

Gas-shielded FCAW, on the flip side, needs that external shielding gas to work properly, remove it, and your weld quality will suffer faster than a burnt toast.

Q2: Can fiber lasers replace CO₂ lasers entirely?

Not quite. Fiber lasers used to be the high-maintenance divas of the laser world, expensive and temperamental compared to CO_2. But those days are mostly over.

Today, fiber lasers have pulled ahead, especially in metalwork. They slice through stainless steel, aluminum, and brass with incredible speed, precision, and almost no maintenance needed.

That said, don't count CO_2 lasers out. They're still the go-to for cutting organic materials like wood, rubber, leather, and acrylic. CO_2 machines deliver crisp, clean edges on plastics and excel at engraving non-metal surfaces.

Think of fiber lasers as the flashy sports car of cutting metal, while CO_2 lasers are the trusty old truck that handles everything else. Each still rules its own lane.

The immortal words of Simon the Supreme Commander.

Q3: How do I know whether to use a CNC router or a CNC laser for a project?

The choice between a CNC router and a CNC laser is a bit like choosing between a chainsaw and a scalpel. It all depends on the material, edge finish, and how complex your project is. CNC routers are like the heavy-duty tool for handling thicker stock, they're the lumberjacks of the CNC world. On the other hand, CNC lasers are the precision surgeons, cutting through thin to medium materials like a scalpel through silk.

Routers tend to leave behind rougher edges with tool marks, kind of like they took the "scenic route" through the material. Lasers, on the other hand, deliver smooth, burnished edges that look like they were polished by a team of perfectionists. Routers are slower when it comes to detailed work, while lasers zoom through intricate cuts like they're in a race, leaving behind little to no cleanup.

Maintenance-wise, routers can be more demanding, those bits wear down quicker than a pair of old sneakers. Lasers, especially fiber lasers, are much easier to maintain with fewer consumables. Plus, routers tend to be louder than a rock concert, while lasers operate so quietly you might start wondering if they're still working.

So, as a general rule: use a router for thick woods, plastics, and soft metals when you need to apply some force, and break out the laser when precision, minimal cleanup, and fine detail are your top priorities.

Q4: What's the best way to quote complex, multi-step projects?

Let's break it down into stages, because who doesn't love a good breakdown?

1. Design/Prep Time
This is where you get to play digital artist. You'll be neck-deep in CAD, nesting, and CAM programming.

2. Machine Time
This is where the machine does all the heavy lifting (literally). Whether you're charging by the hour or flat rate, this is the time where the magic happens. It's like paying for a Netflix subscription, but instead of binge-watching shows, you're binge-watching your machine work its CNC magic.

3. Material Cost
Ah, the good old material cost. Don't forget to include waste and holding stock. It's like buying groceries: sometimes you get exactly what you need, and sometimes, you're left with extra onions that you'll never use, but they're still part of the bill.

4. Post-Processing
This is the fun part, where you get to slap on the finishing touches: welding, grinding, painting, whatever makes your project look chef's kiss perfect. This stage is like getting ready for a night out. It's where you make sure everything shines.

5. Packaging & Logistics
Especially for large parts, because sending a massive piece of metal in a flimsy box is just asking for disaster. This is where your project gets wrapped up (literally) and shipped out. If it's big enough, you might need a forklift and a small army to get it out the door.

And pro tip: keep process notes!
They're like a diary, but instead of confessions, you've got a record of your time and costs. That way, when similar jobs come around, you can pull up your notes and pretend you totally remembered everything you did last time.

...or you can do what I do about 50% of the time: just wing it, toss out a flat rate for the whole project, and let fate decide if you end up broke and eating instant noodles or rolling in cash and upgrading your laser while sipping something bubbly. It's part pricing, part poker, part "eh, let's see what happens." Now you know why I find casinos boring.

Q5: Can I mix CNC processes in one workflow?

Absolutely! If you have the right machinery and skills, combining different CNC processes can create a lot of synergy, and I do it all the time at *CNCROi.com*!

For example, when I built my dining room table, I used a combination of processes for the metal and wood parts.

For the metal base, I:
- plasma cut the metal to form the leg components,
- welded the pieces together using SMAW, and
- sprayed the finished legs with enamel paint.

For the wood top, I:
- CNC routed the surface to flatten it, and
- applied 2-3 coats of tung oil to finish the surface.

Finally, I assembled the metal base to the wood top using wood screws.

Running a custom fab shop? It's chaos, be creative!

Mixing different tools and processes lets you optimize for speed, finish, and material strength.

Just be mindful of tolerances and heat distortion when combining metals with non-metals. It also makes the entire process of custom fabrication a lot more fun.

Q6: Why does nesting matter if I'm not running big batches?

Even a single part can turn into a material-eating monster if it's poorly laid out. It's like trying to fit a sofa into a tiny apartment, you're going to end up with wasted space, frustration, and possibly some tears.

But with smart nesting, you can save material costs, shorten toolpaths (which means less machine time), and keep your operator from throwing a tantrum because they had to flip a sheet for the fifth time.

Think of nesting as "strategic jigsaw puzzle solving", but instead of fun pictures of puppies or sunsets, you're fitting metal pieces together in the most efficient way possible. And just like welding, it's a skill that's best practiced when you don't need it.

That way, when you're in a time crunch and the deadline's staring you down like a hungry lion, you can nest like a pro without breaking a sweat.

Q7: Do I need a cobot to be competitive?

Not today, but possibly soon. While cobots (collaborative robots) are still carrying a bit of a price tag, they come with some serious perks: 24/7 repeatability, fewer rework headaches, and no risk of them calling in sick or demanding vacation days.

Plus, you can scale with demand without needing to hire an extra person, saving you from the nightmare of juggling more salaries. Cobots are a cry-once investment; employees are a constant HR headache, pick your poison.

Many shops test the waters by renting or leasing cobots first, getting a feel for the ROI before taking the full plunge. But here's the kicker: Don't be tempted by those cheap knockoffs.

Sure, they might seem like a bargain, but they have a tendency to break down at the worst possible moment, like right in the middle of a high-profit job, when you're praying for everything to go smoothly.

Cue the endless delays and quality issues. It's like buying a fancy sports car only to find out it has a habit of breaking down... just in time for your road trip.

Q8: What's the future of CNC? Will it all be AI and robots?

The future of manufacturing is augmented, not fully automated, sorry, no robot overlords taking over your job just yet. AI will throw some design suggestions your way, but don't worry, you're still the boss when it comes to saying, "Yeah, that looks good... or nope, try again."

Cobots will take on the boring, repetitive tasks (think of them as the office intern, but much quieter and more efficient), leaving you free to focus on the fun, creative stuff.

Machines will chat with each other through the Internet of Things (IoT), kind of like a nerdy game of robot gossip, but you'll still be the one pulling the strings on how things move along.

You'll remain the mastermind, while your smarter tools do more with less. But here's the catch: be ready to roll with the punches, because this fast-evolving world of manufacturing will throw curveballs your way faster than you can say "machine learning."

Adaptability and foresight will be your best friends, so get ready to embrace the future, just don't expect it to come with an instruction manual.

CNCs can make anything, with enough tools and blind faith.

Q9: How do I handle clients who don't understand custom work?

Use analogies your clients can actually relate to. Instead of saying, "We're using CNC plasma with a 4 gauge plate using an outer offset of 1/16" for kerf width compensation," try something like, "It's like laser-cutting custom metal puzzle pieces from a thick sheet, precisely sized for your needs."

And don't just explain, show them. Share material samples, mockups, or even time-lapse videos.

I do this all the time at *CNCROi.com* using real footage from my shop. It not only proves I can deliver what I quote, but it builds confidence, clients see the process, the quality, and that I know what I'm doing. Even if my price is higher, they're reassured they're not hiring someone who's going to fumble the job, or worse, never complete it to spec.

Bottom line: Education sells. Confused clients don't buy.

Q10: What's the best advice for someone starting their custom CNC business?

Learn by doing, because let's face it, failure often teaches way more than any manual ever could. And, let's be real, sometimes manuals are just fancy paperweights anyway.

Start with one versatile machine, like a CNC router or laser, and slowly build up your empire. Rome wasn't built in a day, and neither was your tool collection.

As you grow, build a killer portfolio to show potential clients exactly what you can do. I've created over a thousand blog posts and videos that work around the clock, attracting customers while I'm busy doing other stuff at the shop or at home. You can do the same, just get yourself a camera and start making some content.

When quoting projects, make sure you value your time. Undervaluing yourself is like giving away free snacks at a party, you're doing all the work, but no one's thanking you for it. And hey, don't chase every trend like a dog after a tennis ball, focus on mastering your core skills first.

You don't need every tool under the sun, just the right ones for the jobs you actually plan to tackle. Unless, of course, you're secretly a tool collector, then by all means, go wild and enjoy the adventure.

Q11: How can I improve part accuracy on my CNC machine?

To achieve higher accuracy in your work, start by making sure your machine is properly calibrated and well-maintained, because let's face it, a machine that's out of whack is like a car with a wonky wheel. It's going to get you somewhere, but probably not in one piece.

Checking tool offsets is key, because if they're off, your cuts will be as consistent as a toddler's mood swings. And don't skimp on materials! Using high-quality stock is like choosing a premium steak over a mystery meat hotdog. Sure, it might cost more, but the results are worth it.

And when you're cutting for hours on end, apply some coolant. It keeps things cool and stable, preventing material expansion and warping, basically, it's like giving your machine a chill pill.

Regular inspections and fine-tuning your settings aren't just for immediate results, they also contribute to long-term accuracy, reliability, and the general feeling of peace when everything works the way it's supposed to.

Q12: What's the best way to handle heat distortion in metals during laser cutting?

Heat distortion is your laser's way of saying, "I'm running too hot." Under intense heat, metal expands and contracts, which can lead to warping and dimensional inaccuracies. These effects are even more pronounced in CNC plasma cutting due to its higher heat input and more aggressive heat-affected zone, but we will stay focused on laser cutting since the same thermal principles apply.

To prevent parts from going sideways, effective cooling is essential. Water-cooled fixtures help control heat buildup, maintain flatness, and preserve dimensional stability.

Staged cutting is another effective strategy. Allowing the material to cool between passes, especially on thicker sections, reduces thermal stress and limits distortion.

Cutting speed also plays a major role. Excessively slow speeds concentrate heat and increase the risk of warping, while overly fast speeds can compromise cut quality or prevent full penetration. Finding balance is critical.

In certain applications, preheating and controlled cooling techniques can further reduce distortion by managing thermal gradients.

Engraving can be even more demanding from a thermal standpoint. Deep etching, particularly with fiber lasers, involves slow, repeated passes that allow heat to accumulate rapidly, increasing the likelihood of distortion.

Q13: Why does my plasma cutter create a rough edge on thicker materials?

Plasma cutting uses an electrical arc to melt through material, which sounds cool in theory, but in practice, it can leave you with rough edges, kind of like trying to carve a turkey with a chainsaw. When working with thicker metals, it's essential to match the amperage to the material thickness. Too much power? You'll end up with edges rougher than your last DIY haircut.

Cut speed is just as crucial, go too fast, and you might as well be using a paper shredder; go too slow, and you'll just create more work for yourself. Finding that Goldilocks speed is key for a better finish.

If you do end up with jagged edges, don't panic, post-processing techniques like grinding or sanding can help smooth things out. And if you're dealing with thicker materials where precision really counts, consider upgrading to a fine-cut plasma torch or, better yet, switch to a waterjet or laser or HD plasma systems for those cleaner, more refined results. Because sometimes, it's better to be a bit more delicate, unless you're cutting a Hulk-sized steel beam, then go big or go home.

Q14: Can I use a CNC router for metal cutting?

Yes, technically, CNC routers can cut metal, though it's not exactly something I've been brave enough to try at *CNCROi.com*. While they're mostly built for softer materials like wood or plastic, with the right setup, they can handle certain metals, if you're feeling adventurous.

To get this metal-cutting party started, you need the right bits, like carbide or coated tools, and a machine sturdy enough to resist vibrating like a caffeinated hummingbird. And don't forget the magic ingredient: *A LOT OF LUBE!!!*

Seriously, a good coating keeps things cool and smooth, like that perfect dance partner who never steps on your toes.

Speed control is crucial too; think of it like driving a car in a school zone. Lower RPMs and slower feed rates are your friends when working with metal. But for thicker metals or higher precision, a CNC plasma or laser cutter is usually your better bet, unless you want your router to turn into a metal-flinging pinball machine.

Of course, there are CNC routers that are built like tanks, specifically for cutting metal.

These bad boys have beefy gantries, misting systems, and all the bells and whistles. But let's be real: they're still not CNC mills, which is still your go-to for metal.

If you're asking whether a regular CNC router can cut stainless or tougher metals, I'm guessing you're not talking about those niche, metal-eating beasts, you mean the regular kind, the ones that flinch at aluminum shavings like a cat at a vacuum cleaner. In that case, I'd say don't bother.

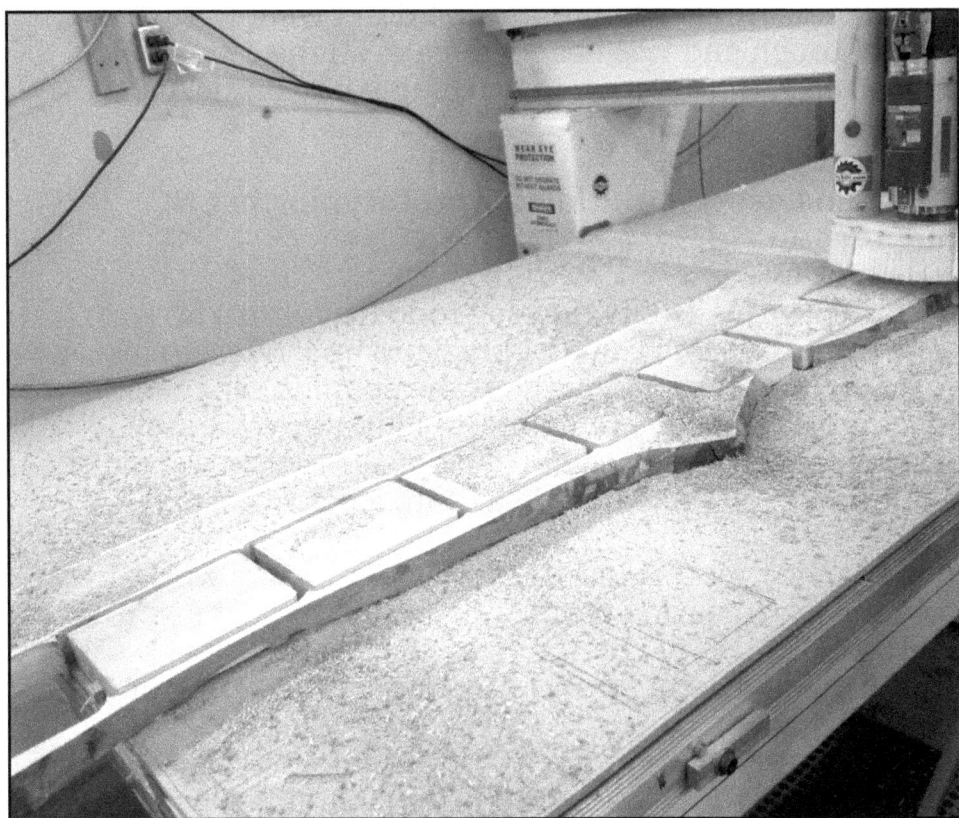

Some machine layouts are more flexible than others.

Q15: How can I prevent material warping during CNC cutting?

Warping is like that awkward moment when your pizza dough doesn't want to stay flat, and your material is no different. It happens when uneven stress distribution or thermal expansion during cutting makes things go all "wobbly." To prevent this, use fixtures to lock that material in place like a bad roommate, no moving allowed.

Cutting from the center outward is also key; it's like spreading peanut butter on bread, start from the middle so the edges don't get all weird and bumpy. And, just like some people are more prone to bad hair days, certain materials are more susceptible to warping. Thicker materials may be more resistant, but they still need some TLC. You can always "pause" to let things cool down as well.

For delicate materials, slow and steady wins the race, think of it like being on a first date. You don't want to rush; a lighter touch and less heat can make all the difference. Sometimes, the solution to warping isn't muscle, it's finesse.

Q16: Can I use 3D printing for prototyping in CNC manufacturing?

Yes, 3D printing is like the fast-food drive-thru of prototyping, it lets you quickly whip up models to test form, fit, and function without the commitment of CNC machining. You can crank out rapid iterations and experiment with those wild geometries that would make a traditional CNC machine break into a cold sweat.

But, like fast food, it has its limits. 3D printing works best for non-load-bearing parts or components that just need to look good and give you the "yep, that works" vibe. It's like trying on clothes before buying, but with less mirror drama.

A smart approach is to use 3D printing to prototype and then step up your game with CNC machining when it's time for the final, serious production parts.

3D printing has come a long way in 12 years. (3D Systems Cube)

Q17: What's the difference between a waterjet and a laser cutter? Are they interchangeable for projects?

Waterjet and laser cutters, two tools, both alike in purpose, but like comparing a squirt gun to a lightsaber. A waterjet is using high-pressure water mixed with abrasive particles to slice through nearly anything, including metals, stone, and ceramics. If it's thick, hard, or delicate, the waterjet's your go-to.

Meanwhile, the laser cutter is the cool, precise surgeon, zapping through metals and plastics with a focused beam of light, delivering clean edges and high speed. Sure, lasers are faster and more precise, but they do create a heat-affected zone (HAZ), which can be problematic for some heat-sensitive materials.

So, in a nutshell: when cutting tough, thick, or sensitive materials, call in the waterjet for its versatility. But when you need speed, precision, and a clean cut on thinner materials, the laser is your best (but not only) option.

Q18: What's the role of CAD in CNC machining?

CAD (Computer-Aided Design) is like the GPS for your CNC machine, without it, your machine's just driving blindfolded. It takes your brilliant ideas and turns them into digital blueprints that your CNC can actually follow via G-Code without the risk of wandering off course.

Want to avoid turning a simple job into a "what went wrong?" saga? CAD lets you run simulations to see how your design might fail before you even turn on the machine.

It's like having a crystal ball, but way more practical, and way less "mystical." And if you're looking to save time and avoid redoing work, CAD is a game-changer. It lets you tweak your designs on the fly, test out new iterations, and avoid that awful moment when you realize you need a redesign halfway through.

So, whether you're learning CAD yourself or hiring someone who knows their way around it, just think of it as investing in your CNC machine's "driver's license" for precision and efficiency.

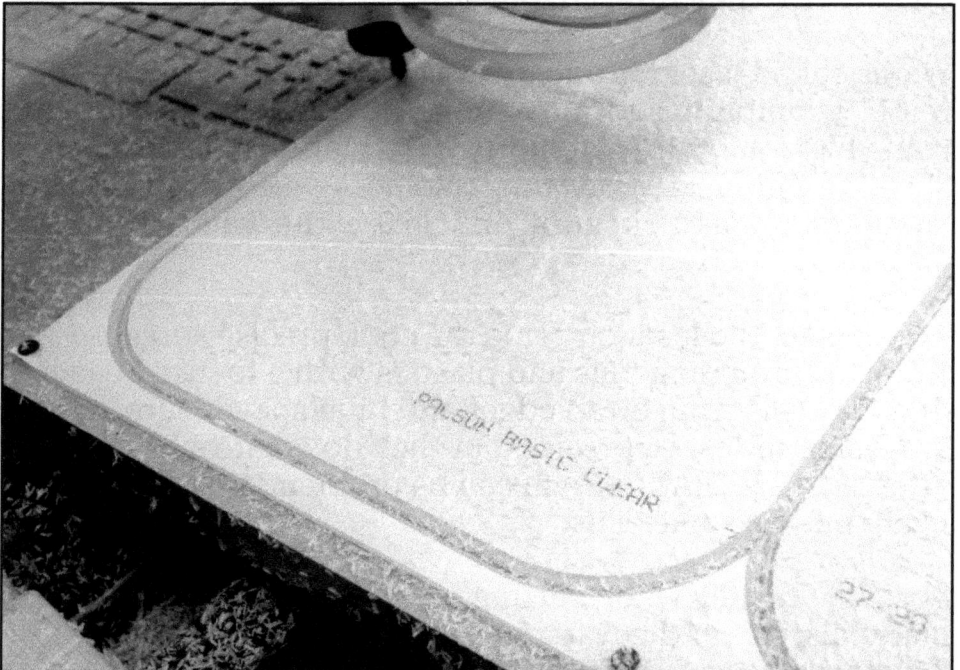

Polycarbonate: rigid and very hard to hold down!

Q19: How do I handle difficult-to-machine materials?

Machining tough materials like titanium or hardened steel can feel like trying to cut through a brick with a plastic spoon, it's just not going to work well without the right strategy. To avoid turning your tools into expensive paperweights, slow and steady wins the race: lower cutting speeds help reduce heat buildup and keep your tools from tapping out early.

Specialized coatings are your new best friend, think of them like sunscreen for your tools, keeping them cool under pressure and reducing friction. As for the geometry of your tools, well, this is where you can't just "wing it." Some materials require tools that are built like bodybuilders, tough enough to withstand the stress without breaking a sweat (or breaking the tool).

And when all else fails and your CNC tools are giving you that "I can't deal with this" look, don't be afraid to turn to alternatives like waterjet or EDM. They're like the "heavy-duty" options for when the going gets tough and your standard CNC tools start asking for a vacation.

Q20: How can I ensure my CNC machines stay efficient?

Efficiency in CNC machining is like making the perfect cup of coffee, it's all about the right balance of ingredients and a bit of TLC. First, let's talk maintenance. Keeping your machine well-lubricated, calibrated, and cleaned is like giving it a spa day, it'll run smoother and live longer, preventing that all-too-familiar "why is it making that noise?" moment.

Next up, toolpath optimization. Think of it like planning your route for a road trip, if you avoid traffic, you get there faster. Efficient programming reduces unnecessary tool changes and shortens production time. And when it comes to materials, don't skimp! Using cheap materials is like trying to build a house with wet spaghetti, it's not going to end well, and your machine will be the one picking up the pieces.

Finally, invest in solid software for toolpath optimization, because your CNC machine deserves more than just a half-baked navigation system.

Keep an eye on performance, and you'll be running a well-oiled, high-performing operation instead of a chaotic "does anyone know why this isn't working?" situation.

Q21: How are PQR and WPS different but also the same in welding?

Think of the PQR (Welding Procedure Qualification Record) as the "proof of the pudding", it's the evidence that the recipe works. You've tested it, you've got the results, and now you can confidently say, "Yep, this welding procedure is solid!" It's like getting a report card after your welding procedure has passed the test (literally).

Meanwhile, the WPS (Welding Procedure Specification) is your welding instruction manual, but without the confusing IKEA diagrams. It's the playbook that tells you exactly what to do, from amperage to travel speed, so you can repeat the perfect weld every time.

It's the difference between hitting a home run and hoping you don't accidentally weld your workpiece to the floor.

Q22: Should I invest in a GTAW welder first, or skip it and go straight to a fiber laser welder?

If you need ultra-clean, precise welds, especially on stainless, aluminum, or exotic metals, and you're working at small to medium scales, fiber laser welding is the future. It's faster, cleaner, and produces far less heat distortion. It's also highly repeatable and often eliminates the need for post-processing like grinding or polishing.

But: fiber laser welders are very expensive. Even the "affordable" models are a major investment. They also have a steeper learning curve, and welding thick sections (>6 mm or 1/4") requires high-end setups that cost even more. Plus, repairs can be slow and pricey.

GTAW (Gas Tungsten Arc Welding), on the other hand:

- Is much more affordable to get into, even top-tier machines.
- Works across a wider range of materials, thicknesses, and weld positions.
- Remains the gold standard for aerospace, food-grade, art, ornamental, and prototype work.
- Is serviceable, you can often fix issues yourself if you're a bit handy.
- Offers unmatched control, you can fine-tune heat input, filler, and arc characteristics with precision.
- Has strong community and support, tons of resources, tutorials, and shared knowledge.
- Doesn't require eye-wateringly expensive PPE or dedicated enclosures like laser systems.

If you're focused on prototyping, repair, or low-volume production, start with GTAW. You'll learn key fundamentals that translate directly to laser welding later.

If you're doing medium- to high-volume sheet metal work and can handle the cost, jump to fiber laser, the speed and quality will pay off fast.

There is also a third, more cost-effective alternative to GTAW and fiber lasers that you may not have considered: oxyacetylene brazing and welding. It takes a bit of practice, but you have incredible flexibility with this tool as well.

Because One Book Can't Cover Everything

There are many questions you have not yet thought of, but experience will bring them to you in time. Do not worry, I have you covered. I wrote two books back to back. This book is one of them, and the other is a Q and A companion book that was released about six months later and goes into much deeper, more specific detail.

See page 350 for more information.

Appendices
Deeper Dives and Handy Extras
Charts, Specs, and Tools to Keep You Smart

Appendix A: Imperial–Metric Conversion Sheet

Whether you're on a CNC, laser cutter, or just squinting at a tape measure wondering why the numbers hate you, this conversion chart is your best friend. It bridges the eternal divide between inches and millimeters, fractions and decimals, imperial chaos and metric logic.

Because nothing slows down a build faster than asking "Wait... how many sixteenths is that again?" while your material smokes on the table.

Whether you're programming offsets, double-checking design specs, or arguing with a stubborn file that only speaks millimeters, this guide will save your sanity, your parts, and possibly your reputation.

One inch equals exactly 25.4 mm. No, it's not a typo. It's a weirdly specific number that's kept machinists up at night for generations. So, that 12" x 24" sheet you've been eyeballing, it's actually 304.8 mm x 609.6 mm, because the metric system doesn't believe in rounding or mercy.

Imperial (inches)	Metric (mm)	Common Use
1/32" (≈ 0.03")	0.5 mm	Fine laser cuts
1/16" (≈ 0.06")	1 mm	Typical anodized aluminum tags
1/8"	3 mm	Common MDF thickness
1/4"	6 mm	Light structural parts
1/2"	12 mm	Heavy-duty signage
1"	25 mm	CNC router spoilboard

Appendix B: Metal Gauge vs. Thickness

Always double-check gauge versus actual thickness, because the metal industry apparently decided that consistency was optional and confusion builds character.

Steel, aluminum, and brass all play by their own rules, like three siblings who grew up in the same house but now speak entirely different dialects. One's metric-ish, another's imperial-adjacent, and the third just does whatever it wants on weekends.

So if you think 16 gauge means the same thing across the board, surprise! It doesn't. In fact, it might be thicker, thinner, or somewhere in between depending on the material, and nobody's sending out a memo.

Welcome to the chaotic neutral of measurement systems, where double-checking is mandatory and blind trust gets expensive.

Imperial (inches)	Gauge	Inches
20 ga	0.91 mm	0.036"
18 ga	1.27 mm	0.050"
16 ga	1.59 mm	0.063" *thinnest I'd cut using my CNC plasma or laser due to HAZ
14 ga	1.90 mm	0.075"
12 ga	2.66 mm	0.105" *mild or stainless steel go-to thickness for tags through signage
10 ga	3.42 mm	0.135"
7 ga	4.55 mm	0.179" *great for firepits
3 ga	6.07 mm	0.239" *SMAW leads and machine grounding

Appendix C: Material Selection Guide

When picking materials for custom projects, looks and cost are just the start. This very rough table breaks down the key stuff, melting point, and how each material likes to be cut.

From un-meltable carbon fiber to metals with moods and plastics that fake looking alike, every material has its quirks. Use this guide to avoid headaches and pick the right tool for the job the first time.

Composite	Notes
Carbon Fiber	*Extremely strong, lightweight, used in aerospace and automotive; typically cuts with CNC Router (beware your lungs) or Waterjet.*

Metals	Notes (Melting Point)
Aluminium	*Lightweight and corrosion-resistant; typically cuts with CNC Router, Fiber Laser, Waterjet, CNC Mill, or CNC Plasma. (≈660°C)*
Brass	*Durable, used in decorative and precision parts; typically cuts with CNC Router, Fiber Laser Waterjet, or CNC Mill. (≈900–940°C)*
Cast Iron	*High wear resistance, brittle; typically cuts with CNC Mill or CNC Plasma. Oxy-Acetylene is commonly used for thicker materials. (≈1150–1300°C)*
Carbon Steel	*Strong, cost-effective, rust-prone without coating; typically cuts with CNC Router, CNC Plasma, Waterjet, or CNC Laser. Oxy-Acetylene also used for thick cuts. (≈1450–1540°C)*
Copper	*Excellent conductivity, soft material; typically cuts with CNC Router or Fiber Laser. (≈1085°C)*

Metals	Notes (Melting Point)
Mild Steel	Great for welding and structural parts; typically cuts with Waterjet, CNC Plasma, or CNC Laser. Oxy-Acetylene is often used for thicker steel. (≈1425–1540°C)
Stainless Steel	Strong, corrosion-resistant, more expensive; typically cuts with CNC Router, Fiber Laser, or CNC Plasma. Oxy-Acetylene is also used for thicker cuts. (≈1400–1450°C)
Titanium	High strength-to-weight ratio, corrosion-resistant; typically cuts with CNC Router, CNC Mill, or Fiber Laser. (≈1660°C)

Plastics	Notes (Melting Point)
Acrylic	Clean edge cuts with CO_2 laser; typically cuts with CNC Router or CNC Laser. (≈160°C (softening))
Delrin (Acetal)	High wear resistance, used in mechanical parts; typically cuts with CNC Router or CNC Mill. (≈175°C)
Polycarbonate	Tough, impact-resistant, used in engineering; typically cuts with CNC Router or CNC Laser. (≈147°C)
Polyethylene (HDPE)	Chemically resistant, used in containers and signage; typically cuts with CNC Router. (≈130–137°C)
Polypropylene	Light, resistant to fatigue and chemicals; typically cuts with CNC Router or CNC Laser. (≈160–170°C)
PVC	Easy to machine, used in pipes and fittings; typically cuts with CNC Router. (≈75°C (softening))

Rubber	Notes (Melting Point)
Silicone Rubber	Flexible, heat-resistant, used in seals; typically cuts with CNC Router or CNC Laser. (≈200°C)

Woods	Notes (Cut with CNC Router & CNC Laser)
Acacia	Dense, durable, used in furniture and flooring
Black Walnut	Dark, strong, used in high-end furniture and cabinetry
HDF (High-Density Fiberboard)	Denser than MDF, good for engraving and cutting
Maple	Strong, fine grain, used in furniture and cabinetry
MDF (Medium-Density Fiberboard)	Uniform density, ideal for routing
Oak	Strong, durable, often used in furniture
Plywood (Birch)	Excellent for engraving & light builds

Other	Notes (Melting Point)
Glass	Brittle but strong, used in precision work; typically cuts with CNC Router (specialized diamond-coated bits) or Waterjet. (≈1400–1600°C)
Ceramic	Extremely hard and heat-resistant; typically cut with Waterjet or CNC Router using specialized bits. (≈1600–2000°C)

Steamboat parts nested PERFECTLY using a hybrid approach.

My CNC laser cut a steamboat from 1/8" (3 mm) MDF, why not?

Appendix D: Chip Load Explained

Chip load is basically how much your tool bites off per revolution.

Too much, and you'll snap bits like pretzels. Too little, and your tool just rubs the material like it's giving it a gentle massage (not helpful). Get it just right, and it's CNC Goldilocks: smoother cuts, longer tool life, and way fewer swear words in the shop.

Chip Load = Feed Rate / (RPM × Number of Cutting Edges)

Parameter	Unit
Feed Rate	Inches/Minute (IPM) or mm/min
RPM (Spindle Speed)	Revolutions/Minute
Number of Cutting Edges	Typically the number of flutes or cutting edges - dimensionless unit
Chip Load	Inches/tooth (IPT) or mm/tooth

Steps to Calculate Chip Load:

1. Enter the Feed Rate (IPM or mm/min): This is the speed at which the tool moves through the material. It is usually given by the machine or determined based on material and tool.

2. Enter the RPM (Revolutions per minute): The speed at which the spindle turns. This is typically controlled by your CNC machine.

3. Enter the Number of Cutting Edges: This refers to the number of cutting edges on your tool. Common numbers are 2, 3, or 4, depending on the type of tool you're using.

4. Calculate the Chip Load: Apply the formula to find the chip load in inches/tooth (IPT) or mm/tooth. This will help you understand how much material is being removed by each cutting edge.

Example Calculation:

Let's say:
- Feed rate: 60 inches per minute
- RPM: 12,000
- Number of cutting edges: 2

Chip Load = Feed Rate ÷ (RPM × Number of Flutes).
Chip Load = 60 ÷ (12,000 × 2) = 0.0025 inches per tooth.

So, each cutting edge removes 0.0025 inches of material per pass. That means each cutting edge is taking a teeny-tiny 0.0025" snack with every spin, just enough to get the job done without choking the tool.

Think of it like your CNC machine on a strict diet: efficient, precise, and not overeating.

Feel free to tweak the table to match your setup, because every machine has its own appetite!

Chip load isn't just for cutting, carving gets a bite too!

Appendix E: Adhesives

When it comes to custom manufacturing, choosing the right adhesive is like finding the right pair of socks, important for comfort, performance, and, well, not making you look like a disaster.

Adhesives are your secret weapon for joining materials that refuse to cooperate with welding or screws.

Wood Adhesives

Wood loves a good PVA (Polyvinyl Acetate) glue. It's like the go-to option, offering strong bonds and quick setting times, making you feel like a professional without the stress. For those tougher projects, try epoxy or urethane-based glues for even more durability.

Composite Materials

Composites, fiberglass, carbon fiber, and the like, are picky. They need an adhesive that sticks as well as your childhood friendships did. Epoxy resin does the trick, holding strong under stress and environmental changes. (Just don't use it to bake cookies, though.)

Metal Adhesives

Metals need something sturdy enough to deal with temperature swings and pressure. Epoxies, urethanes, or even good ol' superglue (yes, it works on metal!) are your best friends here. For mixing materials, find a hybrid adhesive that plays nice with both surfaces.

Mixed Material Adhesives

Bonding wood and metal? Ah, the love story of custom manufacturing. Epoxies with steel-filler additives or structural acrylics are the matchmakers.

These adhesives are versatile and strong enough to bond different materials into a beautiful, permanent relationship.

Choosing the Right Adhesive

Picking an adhesive is like choosing a dog breed, know what you need before committing. Consider:

- **Material Properties:** Is your material clingy (porous) or a smooth talker (non-porous)?
- **Environmental Factors:** Will your bond withstand heat, moisture, or the occasional tantrum (chemicals)?
- **Strength Requirements:** How much load will your bond carry before it calls it quits?
- **Curing Time:** Are you looking for a quick fix or a slow burn? (Patience, my friend.)

More than once, I've tried a few different brands and types of glue for prototyping that told me right away my go-to, not because they all failed, but because only one held when it mattered most. Sometimes the fancy, expensive stuff flakes out under pressure, while the basic tube from the bottom of the toolbox pulls off a miracle.

Glue-ups, big, small, or any wood, always a sticky situation!

Appendix F: It Looked Great in the Slicer...

3D printing is a glorious mix of magic and mayhem. One wrong setting, one grumpy spool, or a slightly chilly bed, and suddenly your masterpiece looks like a melted candle.

3D printing: *the metric system's sneaky way into American homes*, where trial and error is tradition. But they've come a long way since I started working with them 15 years ago. Generally, though, start with the manufacturer, they know your machine best, and sometimes a firmware update fixes everything.

Warping

Are your corners curling up like they're trying to escape your print bed? Welcome to the world of thermal contraction. Your part's cooling unevenly and rebelling.

Fix It:
- Use a heated bed (not just for show).
- Add a brim or raft (aka "insurance policies").
- Use adhesion aids: glue stick, hairspray, or purpose-made sheets.
- Enclose the printer if using ABS or other drama-prone materials.
- Level the bed, don't eyeball it

First Layer Issues

First layer is squished like a pancake (elephant's foot) or floating mid-air like it's auditioning for Cirque du Soleil? Your Z-offset is off or your bed's about as flat as a potato chip.

Fix It:
- Re-level the bed with paper test or a feeler gauge. Trust your tools, not your gut.
- Dial in Z-offset properly.
- Clean the bed, finger oils are the enemy.
- Print slower on the first layer. Let it settle in, like a cat finding its spot.

Stringing & Blobs

Do your print looks like a spaghetti monster got halfway through a sculpting course? It's likely your retraction settings are off, filament's moist, or nozzle temp is too high.

Fix It:
- Increase retraction distance and speed (but not too much or you'll clog).
- Dry your filament. Yes, even if it feels dry. It's lying.
- Lower the nozzle temp slightly.
- Enable "combing" and "wipe" settings if available in your slicer.

Under-Extrusion / Over-Extrusion

See gaps in walls or elephant-sized seams? You're feeding in too little or too much material.

Fix It:
- Check your filament diameter is set correctly (2.85 mm ≠ 1.75 mm).
- Calibrate your extruder steps/mm.
- Make sure the nozzle isn't partially clogged.
- Don't guess flow rates, measure, calibrate, test.

Layer Shifting

If your print starts normal but ends like it was knocked sideways by a ghost, then something moved when it shouldn't. Belt slipped, stepper skipped, or a wire got tired.

Fix It:
- Tighten belts. Loose belts are floppy liars.
- Check stepper motor drivers for overheating.
- Make sure your print isn't bumping into clips, cables, or your hopes and dreams.
- Reduce speed or acceleration if it's throwing shapes too fast.

Prints Not Sticking to Bed

If your print suddenly detaches mid-job and turns into abstract art then you're dealing with poor bed adhesion or prep.

The first layer is everything in 3D printing. If it fails, the rest is just a time-lapse of disappointment.

Fix It:
- Level the bed. Again. You're going to do this a lot.
- Clean the bed, alcohol wipe, not a used napkin.
- Use adhesion helpers: glue stick, PEI sheets, tape.
- Try printing slower and hotter on the first layer.

Nozzle Clogs

Is your printer's acting like it's printing, but nothing's coming out? Most likely you have a partial or full nozzle jam, due to dust, burnt filament, or gremlins.

Fix It:
- Cold pull if it's partially blocked (Nylon works great).
- Heat it up and gently poke it with a needle (not your finger).
- Avoid low-quality filament with mystery ingredients.
- Regularly clean and inspect your nozzle like it's a critical engine part, because it is.

Cracking or Splitting Mid-Way

Do you see vertical cracks in tall prints, especially on the sides? It's because your layers aren't bonding well, usually from cooling too fast.

Fix It:
- Enclose the printer if using ABS or Nylon.
- Increase nozzle temperature slightly.
- Reduce cooling fan speed after the first few layers.
- Print slower so layers stay warm longer.

3D printing makes pill boxes your meds would brag about.

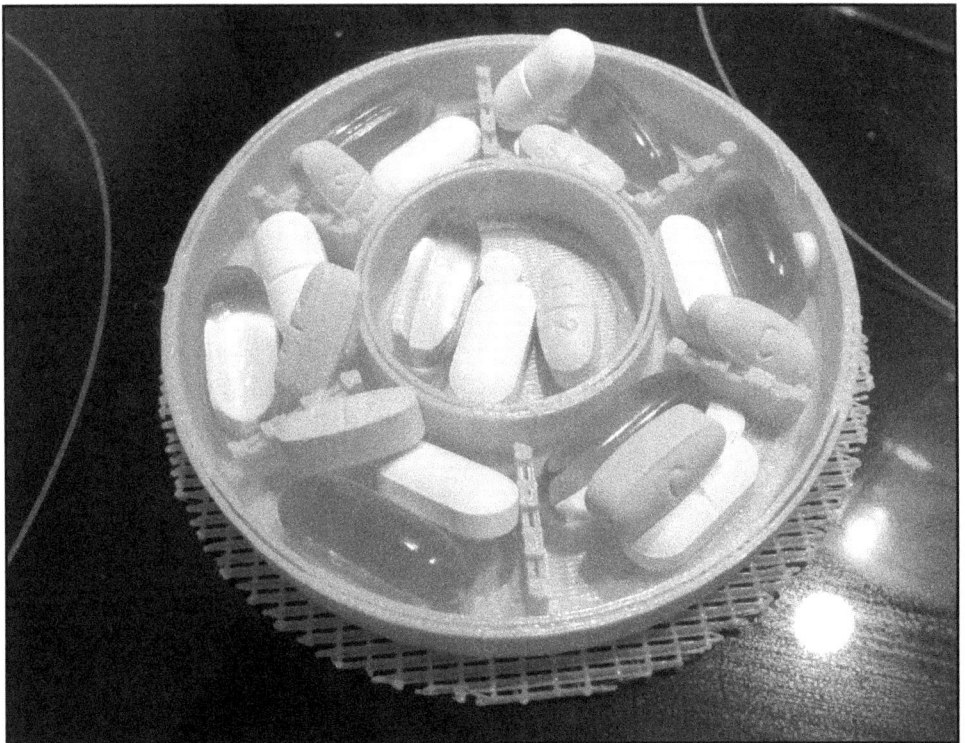

3D printing makes real, functional stuff, not just plastic toys.

Wet Filament (Humidity Soaked)

Hear popping, hissing, or see bubbly surfaces, like your printer's frying bacon? You got moisture in your filament!

Fix It:
- Store filament in airtight containers with desiccant. Tupperware isn't just for leftovers.
- Use a filament dryer
- Bake it (gently): ≈50°C (120°F) for a few hours. Avoid cooking ABS into a LEGO-scented mess.

Dimensional Inaccuracy

If your holes are too small, parts don't fit together or general tolerances laughed at you, you have shrinkage, slicing errors, or firmware calibration is out of whack.

Fix It:
- Calibrate X/Y/Z steps/mm properly.
- Adjust part tolerances in CAD for your material (FDM ≠ injection mold).
- Avoid slicing at low resolution if you're making functional parts.

Thermal Runaway

If your 3D printer stops (hopefully with a warning), the hotend likely triggered a safety shutdown.

Fix It:
- Check thermistor is seated and not damaged.
- Ensure heating element is securely wired.
- Don't bypass this protection unless you really like fires.

Wi-Fi or Cloud Print Failures

If prints fail or lag, it's likely flaky Wi-Fi. Use an SD card or wired connection instead.

Ghosting / Ringing

See echoes or ripples around sharp corners like your printer has the shakes? It's vibrations from moving parts or printing too fast.

Fix It:
- Tighten belts and frame screws.
- Reduce print speed or acceleration.
- Add vibration dampers or heavier base.
- Move the printer off that wobbly IKEA table.

Z-Wobble / Mechanical Wobble

If your prints look like they've spent the night on a wobbly cruise ship, complete with vertical ripples and uneven lines, you've got a classic case of Z-wobble.

This happens when your printer's Z-axis can't move smoothly, leading to artifacts that show up layer after layer. It's not your slicer being creative, it's mechanical mischief at work.

Fix It:
- Check lead screws for straightness and alignment.
- Lubricate the Z-rod (sparingly).
- Make sure your couplers aren't binding or loose.
- Print a test tower to isolate wobble and fine-tune.

Overheating Chamber

Do your parts look like they're melting in slow motion, layer detail blurs, and mechanical parts sag like soft cheese? Fully enclosed printers are great... until they turn into saunas.

Fix It:
- Add ventilation or a temperature-controlled fan.
- Print heat-sensitive filaments (like PLA) with the door open and/or the top cover removed.
- Monitor internal temps if you're running long jobs, your printer isn't a pizza oven.

3D printing a Mayan temple? Now that's history with layer lines.

3D printers: part tool, part toy, all awesome.

Skipped Layers (Sudden Vertical Gaps)

See a horizontal no-man's land appears mid-print? Then the extruder hiccuped or Z-axis ghosted off.

Fix It:
- Make sure filament didn't jam or tangle.
- Check extruder gear for worn teeth or loose grub screws.
- Look for Z-axis binding (or listen, grinding = bad).

Loose Stepper Coupler

If your Z-height is inconsistent or print quality drops after a layer, then that tiny grub screw you ignored backed itself out.

Fix It:
- Tighten your couplers with purpose, not panic.
- Use threadlocker if your printer keeps self-sabotaging.
- Inspect during maintenance, not after it fails.

Dirty or Worn Nozzle

If you are getting inconsistent extrusion or stringy, underwhelming prints, then you may have carbonized gunk buildup, partial clogs, or wear from abrasive filaments.

Fix It:
- Clean nozzles regularly. Don't treat it like a "set and forget" part.
- Replace brass with hardened steel if running anything abrasive.
- Use nozzle cleaning needles, cold pulls, or full replacements if needed.

Power Loss Mid-Print

If your printer stops halfway and doesn't resume then you don't have a "no power-loss recovery feature", or your UPS needs to be replaced with one actually wants to work.

Print Starts in Mid-Air (Nozzle Too High)

If your 3D printer "starts," but it's printing into the void, then your Z-offset was changed, or firmware forgot where it was.

Fix It:
- Reset your Z-offset. Yes, again. It keeps changing like your WiFi password.
- Don't skip homing before prints.
- Ensure probe/sensor is working if using auto-bed leveling.

Thermal Creep

If your filament jams after long prints, especially PLA, you probably have heat creeps up the hotend and softens filament too far up.

Fix It:
- Add or upgrade the hotend's heat break and cooling fan.
- Lower your printing temperature if possible.
- Avoid long retractions if you're pushing filament back into a semi-melted mess.

Printer "Stuttering" on Curves

Noticed your smooth curves coming out looking like they were designed in 1995? If arcs are printing as clunky, low-polygon shapes, your printer isn't interpreting them smoothly. The culprit is usually low G-code resolution or firmware that's not up to the task.

Fix It:
- Enable arc support if your firmware and slicer allow it (G2/G3).
- Use higher resolution exports from your CAD software (STL facet resolution).
- Consider switching to Klipper or a board with higher processing power if you're hitting the limit.

Firmware Gremlins

Get settings randomly reset, print pauses mid-job, or wonky behavior you can't explain? You may have corrupt firmware or loose cables somewhere.

Fix It:
- Reflash firmware.
- Check SD card integrity or swap to another.
- Secure all connectors and inspect wires. Wiggling while printing is a red flag.

Prints Taking Forever

If your 3-hour print mysteriously turns into an 18-hour odyssey. Your layer height is ultra-fine, travel speeds are too slow, or your slicer decided to generate support structures for the air around your model.

Fix It:
- Use coarser layer heights (e.g., 0.2 mm for most prints).
- Increase travel and print speeds within reason.
- Review your slicer settings, do you really need that tree support on a cube?

Inconsistent Extrusion Paths / Wandering Walls

If perimeters don't line up, walls drift or look offset layer by layer then you have loose belts, wobbly wheels, or worn linear bearings.

Fix It:
- Check for frame rigidity and loose components.
- Replace worn V-wheels or linear rails.
- Make sure your nozzle isn't dragging across previous layers (Z-hop, anyone?).

3D printers fail like any good CNC, software bugs, hardware quirks, mid-print spaghetti, and the occasional cursed spool. But don't worry, they're chaotic, hilarious, and when they work, pure magic.

Appendix G: Historical Concise Dives

From the first humans banging rocks together to the laser-precision tech of today, we'll explore the evolutionary timelines of tools, machines, and processes in a very concise form.

Origins of CNC: From Punch Cards to AI

Pre-CNC, Manual Mastery
Before computers, machining was manual. Operators controlled lathes, mills, and saws by hand, mistakes were expensive, and speed was tied to skill (and caffeine).

1940s, Birth of Numerical Control (NC)
In 1949, John T. Parsons and MIT developed punched tape: coded paper strips that guided machines. It was the first step toward automation, replacing guesswork with precision.

1950s–60s, CNC Machines and G-Code
By the '60s, servos and motors could follow tape instructions. G-code programming emerged, creating true CNC systems, huge machines that delivered consistent, programmable results.

1970s–80s, Microprocessors and CAD/CAM
Microprocessors shrank machines and boosted speed. CAD and CAM replaced drafting tables, letting designers send digital blueprints straight to production.

1990s–2000s, Customization Takes Off
Affordable desktop CNCs made routers, laser cutters, and 3D printers accessible. Custom fabrication exploded, with businesses like *CNCROi.com* answering the call for one-off parts and prototypes.

Today, Smart, Connected CNCs
AI, cobots, remote monitoring, and generative design define modern CNCs. From punched tape to smart factories, CNC technology continues to expand!

Welding Through the Ages: From Hammer to Laser

Ancient Roots, The Forge Era
Around 3000 BCE, blacksmiths in Egypt and the Middle East were the original metal artists, fusing metals with heat and hammer. Basically, ancient "DIY, at your own risk."

Middle Ages to Industrial Revolution
Welding was crucial for swords and armor. Then the Industrial Revolution said, "Cool swords, but how about steam engines?", sparking a need for even stronger welds.

1800s, Arc Sparks a New Era
In 1800, Humphry Davy discovered the electric arc, and suddenly welding got a power upgrade. Fire-based methods? So last millennium.

1900s, World Wars and Welding Booms
The 20th century turned welding into a tech race:

> SMAW: Fieldwork hero
> Gas Welding: Great for thin metals
> SAW: Big industry's robotic cousin
> GMAW: Made car manufacturing a breeze
> GTAW: Ultra-precise with moments of zen
> FCAW: Construction's tough guy

2000s–Today, Lasers, Robots, and Magic

> Laser & Electron Beam Welding: Aerospace's secret weapon.
> Friction Stir Welding: Joining metals without melting.
> Robotic Welding: Robots living their best, fully automated lives.
> Fiber Laser Welding: The sleek new favorite for sheet metal and jewelry.

Welding is more than molten metal, it's connection, creativity, and a little bit of controlled chaos.

It's like fusing art and engineering with a lightning bolt while trying not to light your pants on fire.

Lasers: From Lab Invention to Everyday Tool

Lasers started as a twinkle in Einstein's eye in 1917.

It wasn't until 1960 that Theodore Maiman fired up the first one using a ruby crystal, nerdy theory officially turned reality.

1960s–1980s, Lab Coats to Toolbelts
Lasers kicked off in science labs, then quickly made their way into manufacturing, slicing and engraving with way more style (and precision) than old-school methods.

1980s–1990s, Industrial Power-Up
CO_2 and fiber lasers crashed the scene, turbocharging industries like automotive and aerospace. Suddenly, lasers weren't just cool, they were essential.

2000s–Today, Lasers for Everyone
By the 2000s, lasers became affordable, customizable, and a must-have for CNC shops. Fiber lasers ruled, making metalwork clean, fast, and fancy.

Today, Cutting, Welding, and Showing Off
Lasers now handle cutting, engraving, welding, you name it, with crazy accuracy, across everything from aerospace parts to business signs.

Future, Beam Me Up

Precision and efficiency are shaping the future of fabrication.

Laser Welding: Forget slow, clunky welding, laser welding makes joins stronger, faster, and cleaner. It's like a Jedi's precision saber, leaving no mess behind and providing superior, nearly invisible joins.

Additive Manufacturing: 3D printing with metal, layer by layer, using lasers to fuse each section. Stronger than plastic, these parts are practically cast-quality, with laser-precision accuracy that makes each layer bond seamlessly.

Hybrid Systems: Combining lasers with traditional tech gives you the best of both worlds, fast, precise laser work paired with reliable CNC muscle. Efficiency skyrockets while errors plummet.

Micromachining: Tiny, intricate parts for electronics and medical gear? Laser micromachining nails it with microscopic precision. You won't find more detailed work than this without a magnifying glass.

Lasers have evolved from science fiction to manufacturing must-haves. And they're just getting started, the future's looking laser-sharp.

Sadly, steam powered CNCs just never caught on.

Appendix H: Glossary of Terms

Quick guide to custom manufacturing lingo, jargon decoded, and "C4" means Chapter 4 for fast reference.

-#-

2D Profiling
Cookie cutter on steroids, tab or dodge (C4)

3D Contouring
Carves fancy curves, bring patience and snacks (C4)

3D Printing (Additive Manufacturing)
Digital spaghetti machine, genius or glorious blob (C1, C5)

6-Axis Arm
CNC's gymnast buddy, cuts, etches, flexes (C11)

-A-

ABS
Tough plastic, great inside, fades outside (C3)

Ablation
Laser erases layers, pray your material survives (C6)

Abrasive Cleanup
Painful confetti from waterjet fun (C1, C10)

Acrylic (PMMA)
Pretty, brittle, handle like your dreams (C3)

Additive Manufacturing
Stacking layers instead of carving chunks (C5)

Air Assist
Laser's breath mint, crisp cuts, no fires (C6)

Air Cooling
Plasma chill, no rust, less drama (C7)

AI (Artificial Intelligence)
Machine brain, helping, not replacing... yet (C15)

AI-Assisted Workflow
Smarter work, fewer panic attacks (C15)

AI Path Optimization
CNC shortcut magic, cuts like a psychic (C11)

Aluminum
Light, strong, needy, but lovable (C3)

Amperage
Arc muscle, more amps, deeper bites (C7, C8)

Anodizing
Aluminum's spa day, tough, shiny, happy (C13)

Arc
Electric firestarter, and eye regretter (C8)

Articulated Arm
Robot yogi, bends, flexes, builds stuff (C11)

Assembly
Dream meets hammer, good luck (C13)

Assembly Awkwardness
Bad Ikea vibes, awkward parts everywhere (C12)

Assembly Diagrams
Survival maps for confused clients (C14)

Automated Tool Changer (ATC)
CNC butler, swaps tools, saves sanity (C11)

Automatic Nesting
Software Tetris, grain still matters (C9, C12)

A scrap metal desktop organizer, functional and eco-friendly.

A finished metal desktop organizer, sleek, and sturdy.

-B-

Bad Strategy
Cut corners, cry later, do it right (C4)

Beam Quality
Laser's GPA, better beam, happier cuts (C6)

Bead
Molten metal handwriting, messy but glorious (C8)

Bevel Cutting
Angled cuts, prep for welds or flex (C7)

Billable Prototyping
Work = money. Free R&D? Nope (C12)

Bit Breakage
Too fast, too deep, snap city (C4)

Brushing / Graining
Metal haircuts for showroom sparkle (C13)

Bubble Wrap
Fragile part armor, and stress relief (C14)

-C-

CAD (Computer-Aided Design)
Big ideas meet harsh reality (C1, C2, C5)

CAM (Computer-Aided Manufacturing)
Tiny mistakes, big CNC problems (C2, C12)

Carbon Fiber
Feather light, nightmare to cut (C3)

Carbide End Mill
Tougher than your sock collection (C4)

Cast Iron
Reliable, until it cracks dramatically (C3)

Chatter
Vibrations ruining finishes (and moods) (C1)

Chip Load
Balance it, or snap, scorch, sob (C4)

Circular Design
Reuse everything, trash is for quitters (C15)

Climb Milling
Fast cuts, unless you like broken bits (C4)

Clear Coating / Sealing
Sunscreen for parts, stay shiny, stay safe (C13)

Clearance Fit
Wobbly fits = bad handshakes (C2)

Client Approval Checkpoints
"Are we good?" or wallet cries (C12)

Client Handoff
Deliver parts, pray for silence (C14)

CNC Bending
Bend metal fast, no pretzels allowed (C1, C10)

CNC (Computer Numerical Control)
Follows orders, right or hilariously wrong (whole Book)

CNC EDM Wire
Slow lightning cutter, crazy precise (C1, C10)

CNC Lathes
Spin, carve, get dizzy, repeat (C1, C10)

CNC Laser
Cuts anything, except shiny sass metals (C1, C6, C10)

CNC Machining
Metal shaving, straight razor style (C12)

CNC Milling Machines
Shape, grind, slow but gorgeous (C1, C10)

CNC Plasma
Blowtorch + supercharger = thick metal hero (C1, C10)

CNC Punch Press
Metal stapler, fast, basic, no frills (C1, C10)

CNC Router
Dusty, drama-free CNC Swiss Army knife (C4, C10)

CNC Stamping
Cookie-cutter mass production, no fancy stuff (C1, C10)

CNC Waterjet
Water, sand, rage, cut anything clean (C1, C10)

Cobot (Collaborative Robot)
Robot buddy, no drama, no breaks (C1, C11, C15)

Commercial Invoice
Customs' favorite thing to misplace (C14)

Compressed Air
Cheap gas, don't expect miracles (C7)

Compression Bit
Two-faced router hero, top and bottom clean (C4)

Composites
Light, strong, machinist's nightmare fuel (C3)

Conceptual Prototype
3D sketch you can drop (don't) (C12)

Conductive Materials
If it ain't conductive, plasma shrugs (C7)

Conventional Milling
Slow, careful, fragile material approved (C4)

CO_2 Laser
Tortoise laser, wood yes, metal nope (C6, C12)

Crating
Wood armor for parts, forklift wars (C14)

Cut Order
Cut smart, or parts revolt (C9)

Cut Quality
Your edge's A+ or detention (C7)

Cutting Speed
Thin = zoom. Thick = crawl (C6, C7)

Custom Manufacturing
Forget mass production, order gourmet parts (C1)

-D-

Deburring / Edge Cleanup
Save your fingers, ditch razor burrs (C13)

Degrees of Freedom (DOF)
More axes = robot party tricks (C11)

Delrin (Acetal)
Tough, slick, rich kid plastic (C3)

Delta Arm
Tiny robot hand, snatch parts fast (C11)

Deposition Rate
Your weld speedometer, faster = cooler bragging (C8)

Design Assistance (AI)
Smarter than your buddy, no bragging (C15)

Design Brief
Skip it = endless design misery (C1)

Design for Manufacturability
Make it real, not CAD fantasy (C1)

Desiccant Packs
Moisture blockers, don't snack on them (C14)

DFM (Design for Manufacturing)
Dream big, but spare fabricators' tears (C2)

Digital vs. Analog Tools
CNC perfect, human oops, but necessary (C1)

Disassembly-Friendly Design
Think Lego, not crowbar rage (C15)

Dogbones / T-Bones
Router hacks, square peg, round hole fix (C2)

Drais BMT
Bike-born multitool, tiny tank vibes (C12)

Drag Knife
Slow cuts only, rush = wreck (C4)

Drag Knife (Low RPM)
Foam loves chill speeds, no shredding (C4)

Drilling
Vertical holes, peck deep or cry (C4)

Dross
Scrape leftovers, or call it "texture" (C1, C7)

Dry Fit
Test before assembly, regret prevention plan (C13)

Dual-Arm Cobots
Robot jazz hands, work double, no coffee (C11)

Downcut Bit
Clean tops, cranky deep pockets (C4)

-E-

Edge Cleanup
Smooth edges, save your sanity (and fingers) (C13)

Edge Protectors
Foam bodyguards for sharp stuff (C14)

Elasticity
Material yoga, bend, stretch, bounce back (C3)

Electrode
Metal popsicle, fuel for weld dreams (C7, C8)

End Effectors
Robot hands, grip, weld, laser magic (C11)

End Mill (Upcut/Downcut)
Clear chips or clean edges, choose right! (C4)

Engraving
Laser's autograph, burn it, own it (C6, C13)

Engraving Depth
Laser whisper or bonfire, depth matters (C12)

EPS (Encapsulated PostScript)
CNC's BFF, designer's mild headache (C2)

-F-

FCAW
Flux-cored welding, wire with armor, knight style (C8)

FCAW-G
Flux-cored with gas backup, clean welds, gas babysitting (C8)

FCAW-S
Self-shielded flux firepower, no tanks, just sparks (C8)

FDM (Fused Deposition Modeling)
Squirting dreams and sometimes nightmares, layer by layer (C5)

Not glamorous, but someone wanted me to make a UV printer jig.

Feed Speed (RPM)
Spin smart, too fast = fireworks, not cutting (C4)

Fiber Laser
Metal branding iron on steroids, fast, sharp, brutal (C6, C12)

Fiber Laser Welding
Lightsaber welding, clean, fast, slag-free glory (C15)

Fiberglass
Tough, itchy glitter storm, mask up or regret it (C3)

Fine Detail
Tiny cuts so clean, even elves approve (C4, C6, C10)

Finish Durability
If it scratches like a lottery ticket, nope (C12)

Fit (Tolerance Fit)
Too tight, too loose, or just hero-perfect (C2)

Fit Check
Parts fitting right? Or wrestling an elephant? (C13)

Fixture Table
Steel warzone, holes, clamps, zero mercy (C8)

Fixturing
Hold your parts down, or watch them fly (C1, C8, C12)

Flexible Automation
Robots that switch jobs without throwing tantrums (C11)

Focus
Laser sharpness, off by a hair, hello disaster (C6)

Fragile Parts
Shatter dreams in one bad shipment, pack smart (C14)

Frequency
Laser boogies per second, disco or slow dance (C6)

Freight Forwarding
Shipping wizardry, no drama, no lost pallets (C14)

Freight/Pallet Shipping
Big, heavy, forklift fun, and backaches (C14)

Functional Prototype
Rough draft that works... or explodes (C12)

Flux
Magic welding dust, shielding your bead with smoke (C8)

-G-

Galvanized Steel
Zinc-coated steel, cuts fine, fumes like a bonfire. Vent it! (C7)

Gas Type
Your plasma's mood setter, pick wrong, clean longer (C7)

Gauge (Metal)
Thicker metal = lower number. Makes zero sense (C3)

Ghosting
Fuzzy engravings? Tighten mirrors, level the bed (C6)

GMAW
MIG welding, hot glue gun with sparks and gas (C8)

Grain Direction (Wood)
Wood's attitude guide, respect it or suffer warps (C3, C9)

GTAW
TIG welding, molten metal calligraphy at its finest (C8)

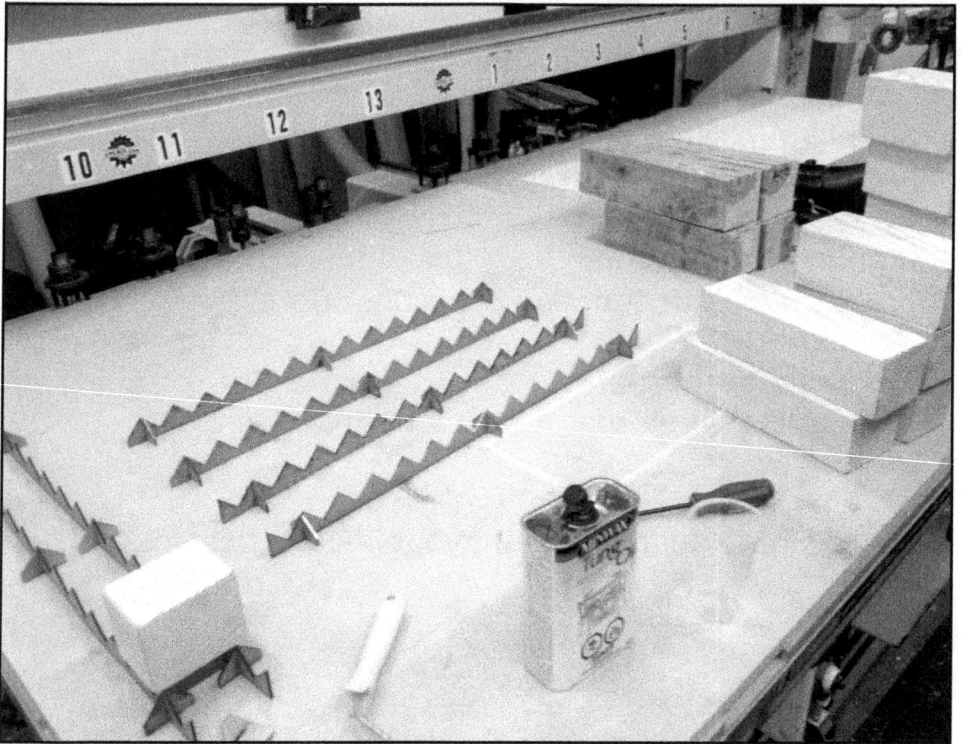

CNC diversity: From firepits to wood blocks, anything goes!

CNC Waterjet: when water cuts what blades fear.

-H-

Hardness
Material's toughness, bring sharp tools or cry (C3)

Hardware Conflicts
When bolts and parts fight like toddlers (C13)

HD Plasma (High-Definition Plasma)
Plasma cutting's fancy cousin, pretty edges, pricey bill (C7)

HDPE / UHMW
Smooth plastic gliders, machine easy, glue dodgers (C3)

Heat-Affected Zone (HAZ)
Edges' sunburn from cutting heat drama (C7, C8, C10)

Heat Distortion
Metal's hot tantrum, goodbye flat, hello wavy (C1, C7, C10)

Hole Misalignment
On-screen perfect, real-life disaster drill lineup (C12)

HSS Tools
Cheap, decent for soft stuff, constant sharpening (C4)

Hybrid Machines
Cut, engrave, print, mill, one machine, all jobs (C15)

Hybrid Nesting
Software starts it, you save the day (C9)

-I-

Improper Cooling
No coolant = meltdown. Metals don't like saunas (C4)

Industrial Automation
Big-scale robots so you can chill more (C11)

Inkjet / UV Print
Pretty prints today, faded memories tomorrow (C13)

Interference Fit
Parts so tight you'll need muscles... or therapy (C2)

International Regulations & Standards
Cross-border paperwork hell, skip it, lose everything (C14)

Ionized Gas
Plasma's secret sauce, fireworks inside your machine (C7, C10)

-J-

Joint
Where metals meet, weld it, don't divorce it (C8)

-K-

Kerf
Cut width reality check, yes, size matters (C1, C2)

-L-

Laser Lens (CO_2/Fiber)
CO_2 loves wood, fiber eats metal, both expensive, don't wreck 'em (C6)

Laser Marking (CO_2)
Surface magic, great contrast, shallow bragging rights (C6, C13)

Laser Marking (Fiber)
Heavy-duty metal engraving & cutting, slow, pricey, but perfect. (C13)

Laser Parameters
Speed, power, frequency, tune it right or torch it all. (C6)

Lead-ins/Lead-outs
Start smart, end clean, skip crash landings (C4)

Layer (CAD)
Color-coded machine to-do list, stay organized, look pro (C2)

Lights-Out Manufacturing
Machines working overnight, you counting money (C11)

Local Drop-Off
Hand-delivering parts, and maybe awkward small talk (C14)

-M-

Machinability
How nicely a material plays with your tools (C3)

Machine Learning
An apprentice that learns but never sleeps (C15)

Magnetic Levitation Rails (Maglev)
CNCs floating like magic carpets, no planets harmed (C11, C15)

Manual Fabrication
Old-school hands-on building, with bonus bandaids (C1)

Manual Nesting
Manually Tetris-ing parts because you don't trust robots (C9, C12)

Manual Plasma Cutter
Handheld torch, fast, messy, loud fun (C7)

Marine Plywood
Waterproof wood superhero, perfect for soggy missions (C3)

Marking
Proof you made it, logos, barcodes, bragging rights (C13)

Mass Production
Make a million copies, no creativity allowed (C10, C11)

Material Flow
Smart part layout = less chaos, more sense (C9)

Material Optimization
Use less material, save more cash (C9)

Material Selection
Pick your material, cheap, strong, pretty... choose two (C1, C12)

Material Shift
When your part moves mid-cut, cue heartbreak (C4)

Material Thickness
How beefy your metal is, know before cutting (C7)

Material Too Dense
Tougher than your tools, pace yourself (C4)

Material Warping
Hot metal curling like angry bacon (C12)

MCAW
Metal-core welding, thick stuff's best friend (C8)

MDF (Medium Density Fiberboard)
Cheap, engraves great, cries in humidity (C3)

Mechanical Engraving
Router marks tougher than your feelings (C13)

MIG Welding
See GMAW, then argue for hours (C8)

MOPA Fiber Laser
Fiber laser's fancy cousin, loves colored metals (C6)

Modular Conveyors
Lego belts for grown-up factories (C11)

Modular Design
Break a part? Swap it, no tears (C15)

Moisture-Sensitive Parts
Materials that sulk in humidity, bag them! (C14)

-N-

Napkin Sketch
Where genius begins, bonus points for mustard stains (C12)

Nesting
Tetris with parts, fit more, waste less (C2, C9, C12)

Nesting Optimization (AI)
Smart AI squeezing parts like a pro (C15)

Nickel Alloys
Tough, pricey, and tool-melting divas (C3)

Not Clearing Chips
Grinding your tool into sadness (C4)

-O-

O-flute End Mill
Cuts acrylic clean, treat it like a diva (C4)

Offset Path
Tool-size cheat, no funhouse mirror parts (C2)

Onion Skinning
Thin layer = no flying parts. Hug it (C4)

Optimized Strategy
Right tools, right settings, no tears (C10)

Optimizing Efficiency
Shave seconds, save dollars, stay sane (C12)

Overdesign
CAD chaos that needs a prayer (C1)

Overengineering
Beautiful, expensive, unnecessarily complicated madness (C1)

Overhead Weld
Sparks, gravity, and shirt fires (C8)

Overpacking
Shipping armor, hurricane-proof your parts (C14)

Oxy-Acetylene Cutting
Old-school firepower, slow but mighty (C7)

-P-

Packaging List
Checklist so nobody loses parts, or their mind (C14)

Pallet Crating
Real shipping armor, tape won't save you (C14)

Pallet Loaders/Unloaders
Robot forklifts, no complaints, no wedding dance moves (C11)

Part Orientation
Tetris your parts right, or suffer later (C9)

Penetration
How deep your weld dives, stick the landing! (C8)

Painting
Turns parts from meh to marvelous (C13)

Plasma
Lightning on payroll, cuts steel like it's paper (C7)

Plasma Cutter
Lightsaber meets blowtorch, cuts metal, fast (C7)

Plasma Cutting
Fast metal haircuts, heat control or bust (C2, C9, C10)

Plasma Table
Where metal gets plasma'd, front-row seat (C7, C10)

Plasma Torch
Fast, fiery metal slicing, easy to mess up (C7)

Pocketing
Tiny router shovel, choose your bit wisely (C4)

Polycarbonate
Tough, clear, pricey, hates lasers (C3)

Porosity
Tiny gas bubbles ruining your weld dreams (C8)

Post-Processing
Cleanup after cutting, more work for plasma (C7, C10)

Power Range
Laser oomph, Hulk smash or delicate whisper (C6)

Pre-Production Prototype
Dress rehearsal for your parts, sparks included (C12)

Predictive Maintenance (AI)
AI stops 3 a.m. machine meltdowns (C15)

Precision
When "close enough" isn't close enough (C10)

Pro Tip
Advice you wish you heard sooner (C9)

Production
Prototype lessons, now on infinite repeat (C12)

Prototyping
Mess up on purpose, fix it later (C2)

Prototyping (again)
Proof first tries are glorified mistakes (C12)

Protective Film Removal
Peeling plastic = CNC Christmas morning (C13)

PVC
Cheap, cuts easy, lasers hate it, though (C3, C6)

Plywood
Strong if good, heartbreak if cheap (C3)

Powder Coating
Metal parts' superhero gym makeover (C13)

-R-

Raster Engraving
Laser cha-cha, great details, slow dance (C6)

Ramp Cuts
Ease in, this ain't a bar brawl (C4)

Rapid Prototyping
Fast ideas, fast failures, pretty comes later (C12)

Real-Time CNC Feedback Loops
CNC tweaking itself like a caffeinated perfectionist (C11)

Refining the Design
Fix the dumb stuff, pretend you planned it (C12)

Reflective Materials
Lasers bounce off shiny divas, plan ahead (C6, C10)

Repeatability
Nail it the same way, every time (C12)

Repeatable Finishes
Good looks, on demand, not by luck (C12)

Repetitive Task Automation
Robots doing boring stuff, no complaints (C11)

Robotic Arm
Never tired, never whiny, perfect coworker (C11)

Robotic Router Head
Michelangelo with a motor and espresso (C11)

Rod Oven
Bake rods, or cookies when boss blinks (C8)

Rough Cutting
Fast, ugly, and good enough, sometimes (C7)

Router (CNC Router)
Beaver with rabies, chews wood, plastics (C2, C12)

Router vs. Spindle
Loud amateur vs. smooth-talking pro (C4)

RPM Too High
Burn tools, ruin cuts, slow down (C4)

Rubbers/Elastomers
Flexible pals, just keep 'em cool (C3)

Where PCB boards go to retire.

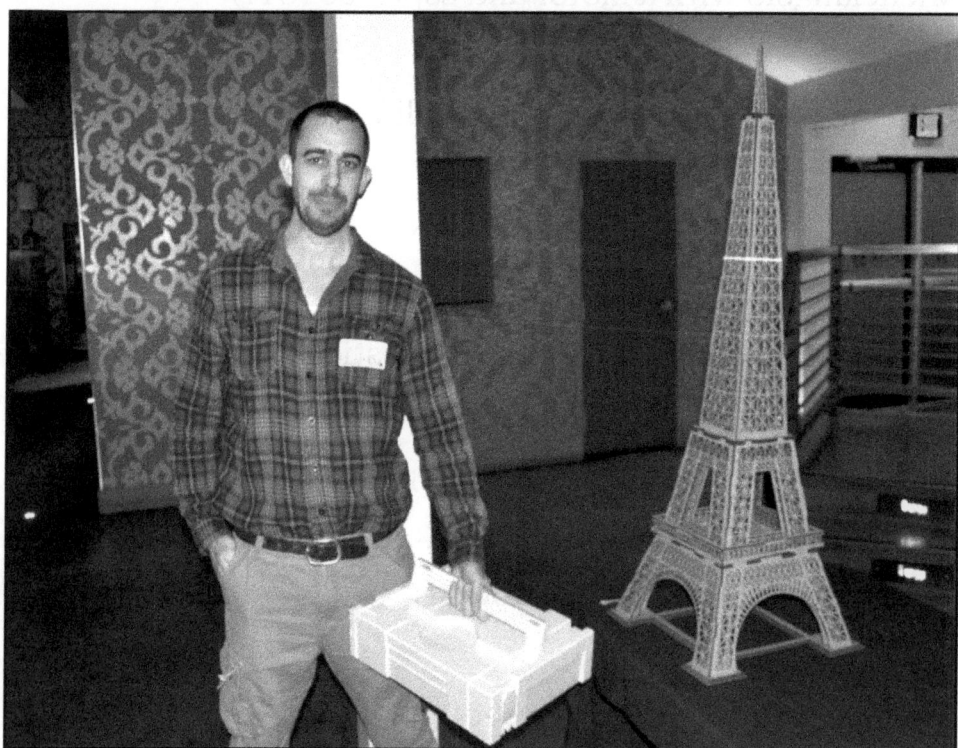

Laser cut Eiffel Tower I designed out of MDF, because why not?

-S-

Scaling
Making more without losing your mind (C12)

SCARA Arm
Stubby robot sprinting side to side (C11)

Self-Healing Polymers
Plastic with Wolverine vibes, scratch, heal, repeat (C15)

Shape Memory Alloys (SMAs)
Metals that magically "remember" their shape (C15)

Signage (Metal)
Metal signs screaming "Look tough!" (C7)

Slot and Tab
Laser-cut Lego magic, if kerf's right (C2)

Softwood
Lumber's eager intern, good for practice (C3)

Spatter
Bad welding's messy fireworks show (C8)

Spindle
CNC's smooth, drama-free power engine (C4)

Spiral Upcut/Compression Bit
Plywood's best friend, clean cuts both ways (C4)

Stainless Steel
Tough, shiny, welding diva metal (C3, C7)

Stamping / Etching
Metal's permanent tattoo, brushes give up (C13)

SMAW
Stick welding, old-school, field-tested toughness (C8)

Shrink Wrap
Plastic film giving parts a hug (C14)

Surface Finishing
Spa day for metal, smooth and shiny (C13)

Sustainability
Building stuff without wrecking Earth (C15)

Synergy
Machines teaming up like Avengers (C10)

-T-

Tabs
Tiny bridges keeping parts from flying (C4, C9)

Tack Weld
Tiny welds, like duct tape, but hotter (C8)

Tamper-Proofing
"Hands off!" seals for nosy people (C14)

Taper
When vertical parts become slanted sadness (C2)

T-Bones (see Dogbones)
Router corner helpers, circles don't square (C2)

Tensile Strength
How much pull before welds scream (C3, C8)

Test Run
First part run, find out what squeals (C12)

Thermal Conductivity
How chill your material stays under heat (C3)

Three-Dimensional (3D) Printing
Layered sculpting, fast, messy, magical (C15)

Throughput
How fast you print cash, or fry machines (C15)

TIG Welding
Steady hands = beautiful molten metal art (C8)

Titanium
Material diva, strong, light, hard to handle (C3)

Tolerance
Define the wiggle room, or physics will (C2)

Tolerance Stacking
Small mistakes = one giant mess (C13)

Tool Life Factors
Treat tools right, or keep buying more (C4)

Tool Path Optimization
Cut faster, cleaner, save time and sanity (C12)

Toolpath
Your CNC's hiking trail, plan wisely (C2)

Toolpath Feedback Loops
CNC's sixth sense mid-cut saves parts (C11)

Toolpath Order
Right cut order = masterpiece, not disaster (C9)

Torch Height
Too high, too low, both bad (C7)

Torch Height Control (THC)
Auto torch magic, cuts stay sharp (C7)

Tube
Metal's stretchy legging, strong and lightweight (C8)

-U-

Undercut
Weld chews where it shouldn't, bad dog! (C8)

UV Laser
Delicate laser surgeon, perfect for fragile stuff (C6)

-V-

V-Bit
Pointy engraver, depth = line magic (C4)

V-Carving
V-bit art, get Z-height right! (C4)

Vector File
VIP passes for CNC brains, no pixels! (C2)

Vector Software
Drawing clean lines like a boss (C12)

Ventilation
Fumes ≠ toughness, air it out! (C7)

Vertical Weld
Uphill/downhill welding mess challenge (C8)

Vision Systems
Robot eyes, because guessing ruins parts (C11)

Visual Inspection
"Does it look right?", trust grandma's method (C13)

Family portraits on Acacia: Faces, meet laser engraved wood.

Family portraits on Acacia: Because wood is the new frame.

-W-

Waste
Scraps = lost cash. Nest smart! (C9)

Waterjet Cutter
Pressure washer with a grudge, no warping (C2, C4, C10, C12)

Waterjet Robotic Control
Robot + waterjet = sci-fi level cuts (C11)

Water Table
Splash zone that cools and calms (C7)

Wear Gloves
Dirty hands = fingerprint crime scene (C13)

Wear-Resistant Prototypes
Parts tough enough to take abuse (C15)

Welding
Bad cuts = weld until it "works" (C9)

Welding Helmet
UV force field for your face (C8)

Welding Symbol
Engineer hieroglyphics, decode or suffer (C8)

Weld Pool
Molten lava steering contest, don't blink (C8)

Wire Feed Speed
Too slow = sputter, too fast = spaghetti (C8)

Wood Packaging Compliance
Bug-free crates = customs survival (C14)

Workflow
Design, cut, finish, dance it right (C1, C9)

Workholding
Hold tight, duct tape's not forever (C4)

-Z-

Z-Axis
Laser's up-down dance, focus or fail (C6)

Zero-Waste Packaging
Save parts, save planet, skip the trash (C14)

Zeroing
Set your start point, or chaos awaits (C4)

Still here? Time to look at my other books! (next page)

Life's a dream, too bad someone else has the remote.

Other Books by Jonathan Cantin

Mastering CNC and Digital Fabrication II
From Curiosity to Competency in 200+ Questions
about CNC Routing, Laser, Plasma, Waterjet, 3D
Printing, and Welding
366 pages / 2026
ISBN 978-1-896369-54-9

Consider this Part II of the book you just finished. It is focused entirely on helping the knowledge from that book truly gel in your mind, while adding additional depth to the concepts and skills you have now acquired.

Together, these two books form a powerful one two combination of CNC knowledge that I wish I had starting over 20 years ago.

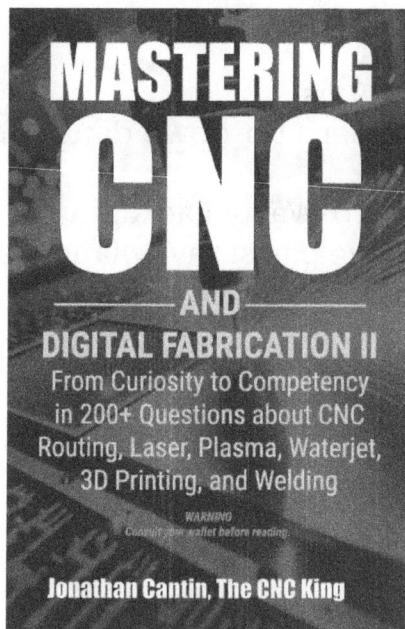

It answers the questions that appear once you are running machines, fixing problems, and moving beyond hobby use.

This is a hands-on reference built around common issues in routers, lasers, plasma, waterjet, 3D printing, welding, and hybrid workflows. You will learn how to improve cut quality, tune feeds and speeds, reduce waste, extend tool life, and solve recurring shop problems with repeatable methods.

Each topic is presented in a quick Q&A format for easy use during setup, production, and troubleshooting.

This book is for makers, manufacturers, engineers, educators, and shop owners who want to understand their machines better and make smarter decisions on the floor.

CNCROi.com V5: Think it, Build it, Sell it!
504 pages / 2015
ISBN 978-1896369525

Two years in the making, this teaches you how to design for CNC machines and successfully market your creations.

This book offers practical advice for designers and business owners, written in an easy-to-understand, non-technical style. It's a great resource for those looking to learn about CNC machines and improve their business.

CNCKing.com Volume 4: Rise of the CNC - Ultimate CNC Design Course
456 pages / 2013
ISBN 978-1896369518

This book covers designing for CNC Table Routers, 3D Printers, and Laser Cutters.

Over 450 pages of projects, expert interviews, and tips to help you get the most from your CNC machines.

Note: Does not include DXF or EPS files.

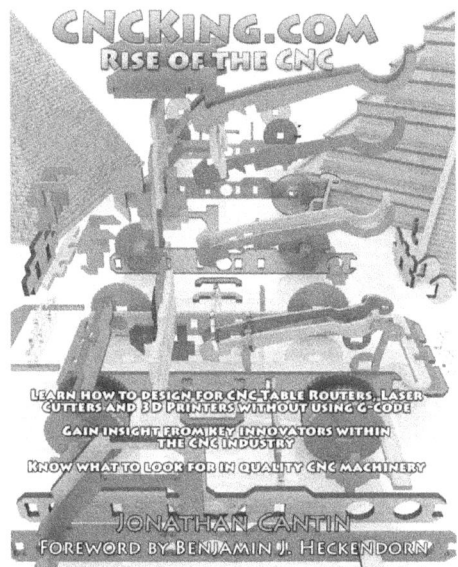

HowToChinese.com: A Practical Mandarin Chinese Course

412 pages / 2012
ISBN 978-1896369501

Learning a new language is fun with a book that guides you every step, written from a student's perspective!

Xiaobei's passion for teaching Mandarin comes to life through her video series, perfect for beginners and advanced learners alike.

Key Features:

Step-by-step process with building chapters

Exercises, real-life dialogues, and mini-dictionaries

Pinyin, Simplified Chinese, and cultural insights

170+ photos from China

Comprehensive video series to boost fluency

Focused on Putonghua, this book offers a complete learning experience, with bonus materials and multimedia support. Written by Xiaobei and Jon, it's the perfect start to mastering Mandarin.

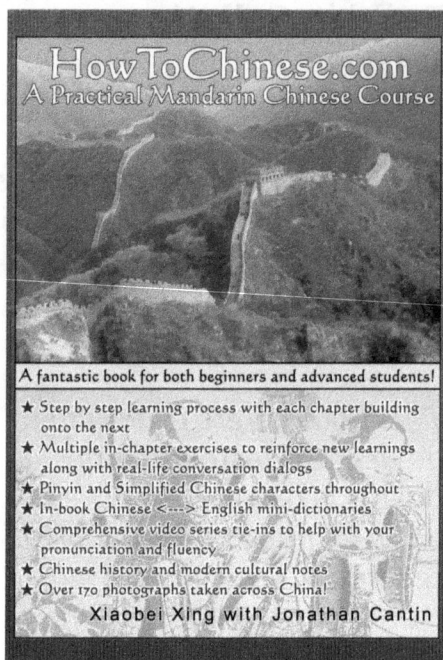

HowToChinese.com
A Practical Mandarin Chinese Course

A fantastic book for both beginners and advanced students!

★ Step by step learning process with each chapter building onto the next
★ Multiple in-chapter exercises to reinforce new learnings along with real-life conversation dialogs
★ Pinyin and Simplified Chinese characters throughout
★ In-book Chinese <---> English mini-dictionaries
★ Comprehensive video series tie-ins to help with your pronunciation and fluency
★ Chinese history and modern cultural notes
★ Over 170 photographs taken across China!

Xiaobei Xing with Jonathan Cantin

WoodMarvels.com: V3 - Evolution of Wooden Designs
200 pages / 2011
ISBN 978-1896369495

This book features woodworking projects across five skill levels with 3D step-by-step assembly instructions. Our unique philosophy: all blueprints are measurement-free, allowing you to scale projects based on wood thickness. Included models from WoodMarvels.com:

Note: Does not include DXF or EPS files.

WoodMarvels.com, Volume 2: Laser Cutting Plans
200 pages / 2009
ISBN 978-1896369464

This WoodMarvels.com book includes blueprints and 3D assembly instructions for models across four levels:

Level 1: Simple creatures and everyday objects
Level 2: Vehicles, functional items, and themed designs
Level 3: Advanced projects with unique concepts
Level 4: Large-scale models and mechanical designs

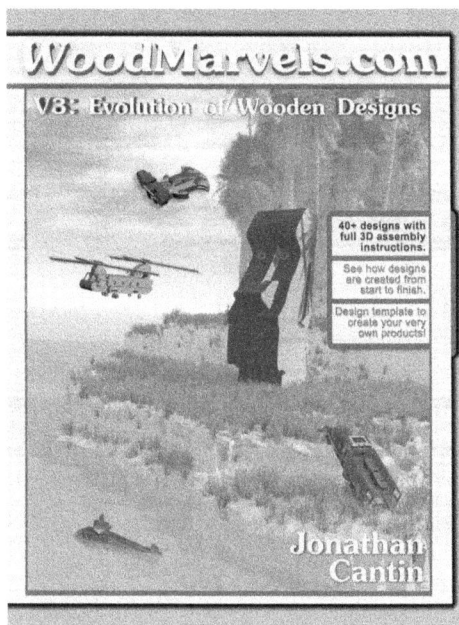

2847 A.D.: Solar Horizons
214 pages / 2009
ISBN 978-1-896369-47-1

In 2847, humanity struggles to survive after near-extinction. Extraterrestrial colonies thrive, and trade is essential.

A Professor and Excur, the most advanced computer, uncover an unexploded neutron sphere in Zulu territory, risking Earth's destruction. Joined by three mining Alabarsi, they seek answers.

What's Project Baptistina, and how is Glass Lake tied to humanity's possible extinction?

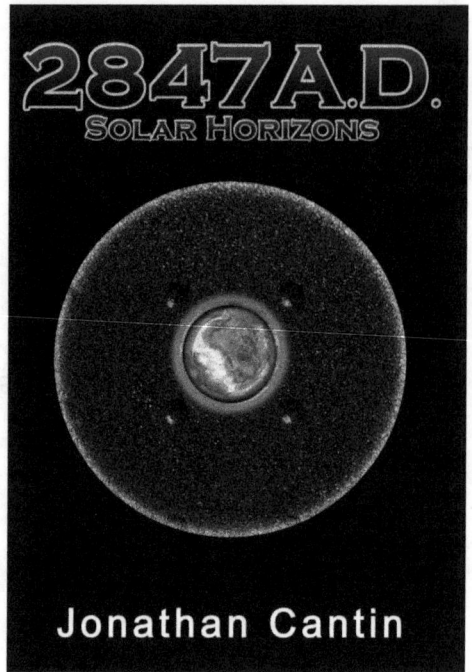

2847 A.D.
SOLAR HORIZONS

Jonathan Cantin

Transforming your life from Blah to Legendary!
140 pages / 2009
ISBN 978-1-896369-45-7

The biggest obstacle to the life you want is yourself.

Creating lasting change requires time, new habits, and a shift in attitude. True success begins with understanding and strengthening yourself.

If you're serious about living a fulfilling life, this book is a great first step. You only have one life, make it count!

TRANSFORMING YOUR LIFE FROM BLAH TO LEGENDARY!

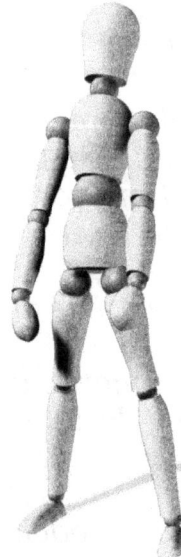

A guide for massive positive and lasting changes in **YOUR** life.

354

www.ingramcontent.com/pod-product-compliance
Lightning Source LLC
Chambersburg PA
CBHW071320210326
41597CB00015B/1285